Praise For
More Bear Cookin'

"PJ Gray offers a playful and affectionate view of American semi-homemade cooking like no other. Our modern lives need easy, delicious, simple recipes, healthful foods, and humor. Mother always said, 'Eat your vegetables.' PJ says, 'Eat your vegetables and don't forget to laugh and enjoy it.'"

—Art Smith, Celebrity Personal Chef and Best-Selling Author
of the James Beard Award-Winning *Back to the Table*

"Hang on, Bears … he's back! PJ Gray keeps his promise with this rich, delicious, and satisfying second helping. This collection *is* bigger and better! There is so much more to enjoy—more recipes, more information, and more humor from this very funny author and food lover. I definitely recommend having this sequel in your cookbook collection."

—Mark Donaway, Chef and Author
of *The Cucina Bella Cookbook*

"PJ Gray's personal touch will warm your heart—and your belly. Pull up a chair at his table; you won't be disappointed. No doubt about it: these recipes are rib-sticking comfort foods. Gray mixes practical tips with glimpses into bear culture and whimsy. The warmth of his personality comes through in his writing, especially in his personal anecdotes about his family and friends. Most of these recipes are appropriate for parties and get-togethers. Gray's multicultural background (French, Irish, and American southern) add to the flavor and variety you'll find here."

—Jonathan Cohen, Author of *Bear Like Me*

NOTES FOR PROFESSIONAL LIBRARIANS
AND LIBRARY USERS

This is an original book title published by Harrington Park Press®, an imprint of The Haworth Press, Inc. Unless otherwise noted in specific chapters with attribution, materials in this book have not been previously published elsewhere in any format or language.

CONSERVATION AND PRESERVATION NOTES

All books published by The Haworth Press, Inc. and its imprints are printed on certified pH neutral, acid-free book grade paper. This paper meets the minimum requirements of American National Standard for Information Sciences—Permanence of Paper for Printed Material, ANSI Z39.48-1984.

MORE BEAR COOKIN'
Bigger and Better

MORE BEAR COOKIN'
Bigger and Better

By PJ Gray

Illustrated by Terry J

CRC Press
Taylor & Francis Group
Boca Raton London New York

CRC Press is an imprint of the
Taylor & Francis Group, an **informa** business

CRC Press
Taylor & Francis Group
6000 Broken Sound Parkway NW, Suite 300
Boca Raton, FL 33487-2742

First issued in hardback 2017

ISBN 13: 978-1-138-42650-4 (hbk)
ISBN 13: 978-1-56023-326-8 (pbk)

This book contains information obtained from authentic and highly regarded sources. Reasonable efforts have been made to publish reliable data and information, but the author and publisher cannot assume responsibility for the validity of all materials or the consequences of their use. The authors and publishers have attempted to trace the copyright holders of all material reproduced in this publication and apologize to copyright holders if permission to publish in this form has not been obtained. If any copyright material has not been acknowledged please write and let us know so we may rectify in any future reprint.

**Visit the Taylor & Francis Web site at
http://www.taylorandfrancis.com**

**and the CRC Press Web site at
http://www.crcpress.com**

Cover design by Terry J.
Cover photo by jasonsmith.com.

Library of Congress Cataloging-in-Publication Data

Gray, P. J.
     More bear cookin' : bigger and better / by PJ Gray ; illustrated by Terry J.
         p.   cm.
     Includes index.
     ISBN-13: 978-1-56023-326-8 (soft : alk. paper)
     ISBN-10: 1-56023-326-5 (soft : alk. paper)
     1. Cookery. I. Title.
     TX714.G7429 2005
     641.5—dc22
                                                                    2005003563

To all the Bears in my life who have inspired me
with their love of food

About the Author

PJ Gray is an award-winning freelance writer and a proud member of the Bear community. PJ authored the popular *Bear Cookin': The Original Guide to Bear Comfort Foods*. He was the former managing editor of *Pride* magazine and is currently a featured writer for *Pink Pages* magazine. He also contributed to the short-story collection, *Rebel Yell 2*. PJ resides in Chicago, Illinois, where he cooks and eats constantly. Learn more about PJ at pjgray.com!

Contents

Section I

LIP SMACKIN' SNACKIN'

Section II
WOOFY BREAKFAST

Section III
MORE HEARTY SIDES

Section IV

COME-AND-GET-IT ENTRÉES

Section V
MORE BEAR MEAT

Section VI
WAY BEYOND THE HONEYPOT

Foreword

It is remarkable how hugely popular the idea of Bears has become. Self-identifying Bears have been around for a mere twenty years, one generation in standard time, but three, if not four, in gay, bi, and queer Bear time.

Although Bears' predilection for food may seem self-evident, it is truly remarkable to see a whole community create its own customs and rituals in pursuit of feasting, frolicking, and communing with food. Few things bring Bears together more than food—preparing, cooking, and sharing with friends. Bears like to eat. As author PJ Gray points out, mealtime is when families sit down together. This assuredly includes Bears' families. Breaking bread together is sustenance—both eating and partaking in the key relationships in our lives.

You have in your hands PJ Gray's second collection of tasty and popular recipes; a how-to manual designed particularly with Bears in mind. It is a sumptuous assemblage of instructions for good times for Bears, from the basic three squares a day to snacks before, during, and after.

How could anyone resist such great food? If there is a Bear cooking in your life (or perhaps that's you), here is the perfect set of instructions—a book full of tasty recipes. Stir and mix well. Serve it up tongue-in-cheek, if you like. Serve it up with the love and warmth that has also been blended in these pages.

Les Wright
Editor, The Bear Book *and* The Bear Book II
Founder, The Bear History Project

Preface

According to my parents, I was appropriately born in the early evening during the dinner hour; and since then, my interest and love for food has never ceased. In fact, my earliest childhood memories were at the dinner table surrounded by my parents and six siblings. I was blessed to have parents who enjoyed cooking and appreciated the effect that dining together had on our family relationships.

As far as we were concerned, no matter who was feuding or despite the family drama du jour, the dinner table was our equilibrium—our neutral ground where the white flags were raised and the weaponry dropped. It was also a place where important news and events were experienced. Report cards were judged, neighborhood gossip was refuted, and adolescent crushes were revealed.

Years later, the announcements became more pivotal and ranged from job promotions, engagements, and pregnancies, to divorces and discarded family secrets. Through all of these events, food was served—hearty, satisfying, flavorful, and comforting. I believe that my parents thought somehow food would make the bad news better and good news even greater. And they were right. They helped define what "comfort food" means to me, and for that I will be eternally thankful.

So much has happened to me since the release of *Bear Cookin'.* The book's promotion allowed me to meet Bears throughout America and from around the world, and I began to share the food I love with a new family. As my Bear Family grew, so did my appreciation and understanding of Bear eating habits. This book is a direct result of that. Somehow, the concept of "food and family" never seems to leave me, it simply reinvents itself.

As with *Bear Cookin'*, I hope readers enjoy this collection of recipes and memories as much as I enjoyed compiling them.

Woof and Big Bear Hugs!

Acknowledgments

I would like to thank my family
for endless love and support.

Special thanks to Bill Palmer and the talented
and dedicated professionals at The Haworth Press.

And extra special thanks
to the worldwide Bear community—Woof!

I would also like to thank the following friends
for their generous support and/or contribution:
John and Stacy Arambages, Susan Arnold, David Cohen,
Mark Donaway, Chicago's Great Lakes Bears, Barbara Gregory,
Janet Hood, Allison Hoste, Bill Keefe, Mary Jo Koizumi,
Rob May, Suzanna McCall, Margaret Resce Milkint,
Jeff Pietrantoni, Catherine Prete, Byron Scott,
Art Smith, Don Stull, Doug Terry,
Greg Tripp, Jon Van Kuiken, and Bill Watters.

Very special thanks to Terry J for his remarkable talent
and generosity. You are a blessing and I am forever grateful.

Thank you!

Section I

LIP SMACKIN' SNACKIN'

HOT CREAMY BEAN DIP

"Hot and creamy"—two words that are certain to attract the attention of any Bear! This has always been a big hit at Bear parties. Once you taste it, you'll know why.

½ cup cream cheese, softened
     (room temperature)
½ cup sour cream
1 (16-ounce) can refried beans
1 tablespoon packaged taco seasoning mix
6 drops hot pepper sauce
2 teaspoons dried parsley flakes
½ cup green onions, chopped
1 cup extra-sharp cheddar cheese, shredded
1 cup Monterey Jack cheese, shredded

Preheat oven to 350 degrees.
Blend in a mixing bowl all ingredients with half of the shredded cheeses until smooth.
Spread mixture into a greased 8-inch square baking dish and top with remaining cheese.
Bake approximately 20 minutes until dip is hot and cheese is bubbling.

Use Your Tool
Best served with strong scoop-shaped corn chips.

More Bearable Meal Suggestions
Ice-cold beer or *New Classic Sangria* (page 19).

2 ½ cups vanilla wafers, crushed
2 cups confectioners' sugar
2 tablespoons cocoa powder
1 cup walnuts, chopped
3 tablespoons light corn syrup
¼ cup brandy

In a mixing bowl, combine the vanilla wafers, cocoa, and 1 cup of the confectioners' sugar thoroughly.
Add the nuts, corn syrup, and brandy and stir well until completely blended.
Form dough into balls (2-inch diameter each) and roll each ball in remaining confectioners' sugar.
Place on a serving platter or store in a sealed container.

NO-BAKE BRANDY BALLS

I love cooking with liquors. When used wisely, they can add extraordinary flavor to the simplest of recipes. Here is a great example. I love serving these during the holidays.

SANTE FE ROLL-UPS

This is an easy way to bring a Southwest flavor to any snacking occasion. You'll love the creamy cheese and corn filling.

1 cup extra-sharp cheddar cheese, shredded
½ cup Pepper Jack cheese, shredded
¼ cup mayonnaise
¼ cup sour cream
¼ cup green onions, finely chopped
1 (11-ounce) can whole kernel corn with
     red and green peppers, drained
10 flour tortillas (6-8 inches in diameter)
10 slices (1 ounce each) fully cooked deli ham

Mix cheese, mayonnaise, sour cream, green onion, and corn.
Top each flat tortilla with one slice of ham.
Spread 2 tablespoons of corn mixture over ham on each tortilla.
Roll up and serve.

More Bearable Meal Suggestions
Beer or *New Classic Sangria* (page 19).

4

1 pound extra-lean ground beef
½ cup red onion, chopped
½ teaspoon garlic powder
1 ½ cups favorite barbeque sauce
1 (10-ounce) package refrigerated
    biscuit dough
1 cup sharp cheddar cheese, shredded
1 cup Pepper Jack cheese, shredded

Preheat oven to 400 degrees.
Lightly grease metal muffin pan (8-cup size).
In a large skillet, cook ground beef, onion, and garlic
powder until meat is browned thoroughly.
Drain grease from pan.
Add barbeque sauce and simmer, stirring occasionally,
for several minutes. Then remove from heat.
Using a rolling pin or the side of a tall, smooth-edged
drinking glass, roll out the biscuit dough on a floured
surface until dough is approximately 6 inches wide.
Place each piece of dough into each muffin cup, folding
the sides into a cup shape.
Fill each cup almost to the top with meat mixture and
top with cheeses.
Bake approximately 15 minutes, or until cheese is
melted and tops are slightly brown.

Use Your Tool
Try replacing your favorite brand of barbeque sauce
with *Your Own Barbeque Sauce* (page 93).

BARBEQUE BISCUIT CUPS

*They taste better
than they look…
sort of like my
last boyfriend.
(Ba-Dum-Ching!)
Thank you, folks,
I'm here all week.
Is this microphone
on?*

CHEESY SAUSAGE BALLS

*Alright,
I'll admit it.
I only serve these
at parties in order
to hear my guests
say, "I just love
your balls."*

2 pounds (2 tubes) uncooked ground pork sausage
16 ounces (or 4 cups) sharp cheddar cheese, shredded
2 cups all-purpose baking/biscuit mix
1/2 cup celery, finely chopped
1/4 cup green onion, finely chopped
1/4 cup yellow onion, finely chopped
1/2 teaspoon garlic powder

Preheat oven to 375 degrees.
Mix all ingredients thoroughly.
Form mixture into 1-inch balls.
Place balls on an ungreased baking sheet and bake 15 minutes or until golden brown.

Use Your Tool
Uncooked balls can be frozen for future use. Simply thaw and follow baking directions.

1 (18-ounce) package frozen mashed potatoes
1 cup whole milk
½ cup ranch salad dressing
2 plum tomatoes, chopped
½ cup red, orange and/or yellow bell peppers,
     chopped
¼ cup green olives, pitted and chopped
¼ cup red onion, chopped
½ teaspoon Worcestershire sauce
¼ teaspoon salt
¼ teaspoon freshly ground black pepper

Prepare potatoes according to package directions
(substitute milk for water if directions require liquid).
Add remaining ingredients to potatoes and stir.
Cover mixture and chill at least 2 hours before serving.

MASHED POTATO DIP

*This is one of
my favorite dips
because it doesn't
compete with
the flavor of the
foods used to dip
into it. In fact,
it enhances their
flavor. Try it with
garlic bagel chips,
flavored tortilla
chips, or veggies
like sliced carrots
and celery.*

A TRIBUTE
Peanut Butter

One day, when I was very young, my mother made me a peanut butter sandwich for lunch and told me this story.

My mother, as a child, lived in France during World War II. Both American and Nazi soldiers occupied her home during the course of the war. Her memories of that time were both fond and sinister. She had little to say about the Nazis, perhaps a defense mechanism in her mind. Perhaps the memories were too dark to share with her child.

Her stories of the Americans, on the other hand, were quite vivid. She recalled happy moments when the charming GIs offered candy and sweets to her and her siblings to the constant dismay of her parents. It was the first time she had ever tasted foods like bubble gum, Hershey's chocolate bars, and most important, peanut butter.

It was the peanut butter, she remembered fondly, that made the strongest impression. She fell in love with it from the first glorious spoonful, and refused to share it with anyone else.

Years later, she married my father and moved to America to become a wife and mother. And, as I recall, we always had peanut butter in the kitchen. Always.

This is what peanut butter means to me.

HOMEMADE CANDY BARS

Candy lover or not, you'll appreciate the fresh taste of these versus a store-bought version.

1 cup creamy peanut butter
1 cup light corn syrup
$\frac{1}{2}$ cup brown sugar, packed
$\frac{1}{2}$ cup sugar
6 cups cornflakes cereal
1 cup semisweet chocolate chips
1 cup peanuts

In a large saucepan over medium heat, combine the peanut butter, corn syrup, and both sugars.
Cook until smooth and well blended, stirring occasionally.
Remove from heat and immediately add the cornflakes, chocolate chips, and peanuts.
Stir mixture until evenly coated.
Pour entire mixture into a greased 9 x 13-inch baking dish.
Press mixture gently and evenly into dish and allow to cool completely before cutting into bars.

More Bearable Meal Suggestions
A tall glass of cold milk.

10

Crust

1 1/2 cups all-purpose flour
2/3 cup brown sugar, firmly packed
1/2 teaspoon baking powder
1/2 teaspoon salt
1/4 teaspoon baking soda
1/2 cup butter, softened
1 teaspoon vanilla extract
2 egg yolks
3 cups miniature marshmallows

Topping

2/3 cup light corn syrup
1/4 cup butter
2 teaspoons vanilla extract
1 (10-ounce) package butterscotch chips
2 cups crisp rice cereal
2 cups pecan halves, chopped

Preheat oven to 350 degrees.
In a large bowl, combine all crust ingredients except marshmallows until texture is crumbly. Press crust mixture firmly in the bottom of an ungreased 9 x 13-inch baking pan.
Bake for 12-15 minutes until light golden brown.
Remove crust from oven and immediately sprinkle with marshmallows. Place pan back into oven and bake an additional 1-2 minutes until marshmallows begin to puff. Remove from oven to cool.
Meanwhile, in a large saucepan over medium-low heat, combine all topping ingredients except cereal and pecans. Heat until mixture is smooth, stirring constantly. Remove from heat and add cereal and pecans. Pour topping over marshmallow crust and spread to cover evenly.
Refrigerate 1 hour or until firm, then cut into bars.

Use Your Tool

A countertop or hand mixer is ideal for mixing the crust ingredients.

BUTTER NUT CHEWS

If you like the Homemade Candy Bars, you'll love these! They're sure to satisfy your sweet tooth.

11

CHEESE CRUD

*Occasionally
I find a family
recipe name that
defies reason.
How and why this
delicious snack
earned this name
is beyond me.
However, as a
child, it made me
laugh every time it
was mentioned.*

13 ounces extra-sharp cheddar cheese, grated
1 cup butter, softened
1/4 teaspoon each cayenne pepper, salt,
    onion powder
3 cups all-purpose flour

Combine all ingredients except flour.
Add flour and mix thoroughly.
With your hands, form mixture into three or four rolls
(each 1-inch in diameter).
Wrap each roll in aluminum foil and refrigerate several
hours or preferably overnight.
When ready to bake, preheat oven to 350 degrees.
Cut rolls into 1/2-inch slices and place on an ungreased
baking sheet.
Bake for approximately 15 minutes or until edges are
lightly brown.
Serve warm or room temperature.

Use Your Tool
An electric or hand mixer is ideal for mixing the
ingredients. Add finely chopped nuts to the mixture for
a twist!

More Bearable Meal Suggestions
Beer.

8 whole kosher dill pickles
2 packages (8 ounces each) cream cheese,
    softened
3 tablespoons garlic powder
3 ounces deli roast beef, thinly sliced

Pat pickles dry with paper towels.
Spread cream cheese around each pickle and sprinkle
lightly with garlic powder.
Wraps two sheets of meat around each pickle and place
them in a sealed container.
Refrigerate for several hours (preferably overnight).
Slice each pickle width-wise to your liking prior to
serving.

Use Your Tool
Enough said.

More Bearable Meal Suggestions
Beer or a dry, very smooth martini.

PC PICKLE WRAPS

Listen, if you're going to "practice safe" with your favorite pickle (and everyone should), then I think this is the best way to go.

13

CONFETTI DIP

As I mentioned in Bear Cookin', *there's an expression that my friends and I use to describe munchies that are so addictive you simply can't stop eating them. We call them "crack snacks," and I assure you this recipe will have your party guests hooked!*

1 teaspoon salt
½ teaspoon freshly ground black pepper
1 tablespoon water
¾ cup cider vinegar
½ cup vegetable oil
1 cup sugar
1 (15-ounce) can each black-eyed peas, pinto beans, and corn (preferably shoe peg), rinsed and drained
1 (4-ounce) jar pimento, chopped or diced
1 medium-sized red onion, diced
1 cup celery, chopped
1 cup green bell pepper, chopped
2 fresh jalapeño peppers, finely chopped

In a large saucepan or stockpot over medium-high heat, combine salt, pepper, water, vinegar, oil, and sugar. Bring to a quick boil stirring constantly until the sugar is melted.
Remove from heat immediately and add remaining ingredients.
Blend and pour into a large covered container and refrigerate for 2-3 hours (preferably overnight so flavor intensifies).
Drain mixture before serving.

Use Your Tool
Warning—a strong, scoop-shaped corn chip is recommended for dipping.

More Bearable Meal Suggestions
Beer or *New Classic Sangria* (page 19).

14

2 pounds extra-lean ground beef
2 eggs
2 tablespoons green onion, minced
1 tablespoon fresh garlic, minced
1 tablespoon lime juice
2 tablespoons soy sauce
12 ounces chili sauce
1/3 cup ketchup
1/4 teaspoon freshly ground black pepper
1 cup cornflakes, crumbled
1/3 cup fresh parsley, minced
16 ounces whole berry cranberry sauce
1 tablespoon brown sugar

Preheat oven to 475 degrees.
In a large mixing bowl, combine all ingredients thoroughly except the cranberry sauce and brown sugar.
Form meat mixture into 1-inch balls.
Place meatballs into an ungreased 15 x 10 x 1-inch jellyroll pan.
Bake uncovered for 10-12 minutes.
Turn oven temperature to low or 200 degrees.
Drain meatballs and either return pan (covered with aluminum foil) to oven or place meatballs into a chafing dish to keep warm.
In a saucepan over medium heat, combine cranberry sauce and brown sugar, stirring occasionally until mixture begins to bubble.
Pour over meatballs and serve in pan or chafing dish.

Use Your Tool
For practical and aesthetic reasons, I recommend a chafing dish for serving.

SWEET CUBBY BALLS

One bite of these sensational balls and you'll know where the expression "sweet meat" originated! This is a more interesting version of the standard party meatball.

MINI HOT BROWNS

It's a snack that eats like a meal. Actually, if you eat it alone, you have a meal!

¾ cup half-and-half
¼ cup hot water
1 chicken bouillon cube
3 tablespoons butter
2 tablespoons all-purpose flour
1 cup Swiss cheese, grated
1 large white or yellow onion, thinly sliced
6 ounces turkey, ham, or roast beef,
     cooked and thinly sliced
6 bacon strips, cooked and crumbled
1 cup green onions, finely chopped
18 slices fresh Italian or French bread

Preheat oven to 350 degrees.
Dissolve bouillon cube in hot water and set aside.
In a saucepan over medium heat, add butter to half-and-half and blend until butter melts.
Whisk flour into mixture until it becomes frothy and the taste of raw flour is removed.
Add bouillon water to mixture, stirring constantly with the whisk until sauce thickens and bubbles.
Add grated cheese to sauce and stir until cheese melts and sauce continues to thicken. (Add additional water and heat if sauce needs to be thinned.)
To assemble, place meat and onion on bread slices.
Top each slice with sauce and crumbled bacon.
Place slices on an ungreased baking sheet and bake for approximately 10 minutes.
Garnish with green onion and serve.

Use Your Tool
If time is not your friend, replace the sauce directions with a jar of heated meat gravy.

1 package chocolate fudge cake mix
1 package (large box) chocolate instant
    pudding
3/4 cup water
4 eggs, beaten
1/2 cup vegetable oil
1/2 cup mini semisweet chocolate chips,
    frozen
1/2 cup dark chocolate chips, frozen
1/4 teaspoon almond extract

Preheat oven to 350 degrees.
In a large mixing bowl, combine all ingredients except
chocolate chips; blend until batter is smooth.
Add chips and blend thoroughly.
Place paper muffin cups in an ungreased muffin pan.
Fill muffin cups three-fourths full with batter and bake
for approximately 30 minutes.

Use Your Tool
An electric or hand mixer is ideal for mixing the batter.

ULTIMATE CHOCOLATE FUDGE MUFFINS

In a world full of designer chocolate and fancy chocolate drink concoctions, these muffins stand as a proud reminder of a simpler time— when the smell of warm, sweet chocolate filled a kitchen and turned a house into a home.

LOVE MY SALSA

Salsa is another type of recipe I love because it is so versatile and easy to make. Besides, no store-bought salsa compares to fresh in my opinion. I've tried so many wonderful versions, but I like this one because of its simplicity and flavor. Experiment and have fun!

8 small tomatoes, seeded and chopped
1 large white onion, chopped
2 stalks green onion, finely chopped
½ cup minced parsley
1 teaspoon salt
Freshly ground black pepper to taste

Toss all ingredients in a large mixing bowl and serve. If serving as a dip, keep at room temperature.

More Bearable Meal Suggestions
Serve as a dip with strong, scoop-shaped corn chips; or spoon over grilled chicken or fish.

18

1 (1.75 liter) bottle dry red wine
4 cups orange juice
1 cup Cointreau, Grand Marnier, or triple sec
2 large oranges, sliced
1 lime, sliced
1/2 lemon, sliced
1 peach, seeded and sliced

Combine all ingredients in a large pitcher and refrigerate for several hours (preferably overnight).

Use Your Tool
It is neither necessary nor recommended that you use an expensive wine. A simple table wine works best. The key with this recipe is to let the fruit do most of the work.

NEW CLASSIC SANGRIA

I fondly recall when I first began to perfect this recipe years ago with my friend Doug while visiting his home in the Napa Valley. By the time we "perfected" it, I awoke days later under the bed of a large Mexican Bear named "Bobo." Don't ask.

19

LA-DEE-DA
APPETIZERS

*In the grand Bear
tradition, this
recipe offers you
the chance to
take something
as simple and
healthy as a water
chestnut and
blanket it with
meaty decadence.
"What could be
more wholesome
or natural?"*
—Auntie Mame

1 package (1 pound) sliced lean pork
   or turkey bacon
2 (8-ounce) cans whole water chestnuts
½ cup chili sauce
½ cup brown sugar
½ cup Miracle Whip salad dressing

Preheat oven to 350 degrees.
Cut bacon slices in half width-wise.
Wrap each water chestnut with a bacon slice and secure
with a wooden toothpick.
Place wrapped pieces on ungreased baking sheet and
bake for 20 minutes.
Meanwhile, in a saucepan over medium-low heat, blend
the chili sauce, brown sugar, and salad dressing until
smooth.
Remove water chestnuts from oven and place in a
9 x 13-inch baking dish.
Pour sauce mixture evenly over pieces and return to
oven to bake for an additional 30 minutes.

Use Your Tool
Remember to use wooden toothpicks when baking.
Avoid any with colored plastic fringe on tops.

More Bearable Meal Suggestion
Beer or a dry, very smooth vodka martini.

1 (12-ounce) jar pickled mixed vegetables,
    drained and finely chopped
¼ cup pimiento-stuffed olives, chopped
2 ounces salami slices, finely chopped
1 tablespoon olive oil
1 tablespoon garlic, minced
2 (10-ounce) cans refrigerated
    flaky-style biscuits
¼ cup provolone cheese, finely shredded
¼ cup sharp cheddar cheese, finely shredded

Preheat oven according to biscuit package directions.
In a mixing bowl, combine vegetables, olives, salami,
garlic and oil. Cover bowl and chill for at least 1 hour.
Bake biscuits according to package directions.
Remove and cool slightly.
Scoop out center of all biscuits being careful to leave
bottom and sides intact.
Stir cheeses into vegetable mixture and spoon
1 tablespoon of mixture into each biscuit.
Return biscuits to oven and bake at 400 degrees for
10-12 minutes.

Use Your Tool
For an added twist, before hollowing out each biscuit,
carefully slice the top off first. After baking, place the
top back on the biscuits before serving.

MINI MUFFULETTAS

*A specialty of
New Orleans,
the muffuletta
sandwich
originated nearly
a century ago and
can now be found
anywhere that you
can get a good
ten-inch round
loaf of Italian
bread. This easy,
mini version uses
biscuits instead
while still capturing
the sandwich's
original flavor,
and it makes an
unusual party hors
d'oeuvre.*

21

BACHELOR BEAR TOAST

Most of us think of corned beef only in hash. This recipe allows you the chance to experience its true delicious nature. Notice the rare occasion that I recommend American cheese over cheddar. In this case, it simply melts better.

8 slices sandwich bread
8 tablespoons butter
4 teaspoons mayonnaise
1 (15-ounce) can corned beef, sliced
Salt and freshly ground black pepper to taste
1 teaspoon dried parsley flakes
4 slices American cheese

Butter each slice of bread on one side only.
Apply mayonnaise to other side of first bread slice, and place in a large skillet over medium heat, buttered side down.
Top bread with a slice of corned beef, a dash of salt and pepper, and a pinch of parsley.
Add slice of cheese on top and a second slice of bread, buttered side up.
Cook for 5 minutes on each side until toast is golden brown and cheese is melted.
Repeat to make three additional sandwiches.

More Bearable Meal Suggestions
French fries or potato chips and a hearty ale.

14 ounces chocolate-flavored sweetened
       condensed milk
1 cup vanilla ice cream
2 candy bars (favorite brand), chopped
1/2 cup pecans, chopped (optional)
2 cups ice, crushed

Add all ingredients to a blender container.
Blend until smooth and serve immediately.

Use Your Tool
WARNING—Repeated smoothie making is inevitable.
Make sure you invest in a well-built, preferably
professional-quality blender for a long kitchen life.

More Bearable Meal Suggestions
Keep your favorite Bear nearby. Drinking this is sure to
make you want to Bear hug!

CANDY BAR SMOOTHIE

*I've had my share
of fruit smoothies.
However, my heart
and Bear belly
belong to this one!
Something this
good shouldn't
be this easy to
make. I suggest
that you keep
these ingredients
stocked in your
kitchen at all
times.*

BUFFALO ROLLS

One bite of these heavenly snack rolls and I guarantee that you'll include them on your "Top Ten Favorite Things I Put in My Mouth List." What? You've never made a list? Tsk-tsk.

19 ounces premade pizza dough
¾ cup mayonnaise
3 ounces bleu cheese, crumbled
4 tablespoons precooked bacon bits
¼ cup fresh parsley, chopped
3 ounces Parmesan cheese, grated
Freshly ground black pepper to taste
¼ cup olive oil
2 tablespoons minced garlic
½ cup (1 stick) butter

Preheat oven to 350 degrees.
Roll pizza dough into a large rectangle.
Coat the entire side of dough facing up with mayonnaise.
Sprinkle bleu cheese crumbs and bacon bits evenly over mayonnaise.
Continue to top dough with parsley and Parmesan cheese and pepper, and roll dough carefully into tube shape.
Cut dough into 12 rolls approximately 1 inch in width.
Place rolls on a baking sheet and allow them to rise for 45 minutes.
Then place rolls into oven and bake for 8-10 minutes.
While the rolls are baking, combine oil, garlic, and butter in a saucepan over medium-low heat.
Simmer for 5 minutes, stirring occasionally.
Remove rolls from oven and brush tops with butter mixture before serving.

Use Your Tool
If you don't own a chef's brush, buy one. If not, try to lightly drizzle butter onto rolls with a teaspoon.

1 large loaf French bread,
  sliced in half horizontally
½ cup butter, softened
1 cup fresh Parmesan cheese, grated
½ cup Monterey Jack cheese, shredded
½ cup Swiss cheese, shredded
1 cup mayonnaise
1 tablespoon fresh garlic, minced
1 bunch green onion, chopped

Preheat oven to 350 degrees.
In a mixing bowl, blend butter, cheeses, mayonnaise,
garlic and green onion.
Spread mixture onto cut side of bread halves.
Bake for 7-10 minutes.
For an added crunch, remove from oven and place
under a broiler for several additional minutes.
Cut into slices and serve.

TUXEDO BREAD

*This bread is
all dressed up
and nowhere to
go…but to your
tummy. This
versatile bread
is also a great
accompaniment to
meat casseroles.*

CARLY'S CLAM PIE

This yummy snack is dedicated to all of my Bear friends in New England. It was inspired by a visit with my friend Carly, a longtime resident of Martha's Vineyard. Maybe you know her?

2 (6 ounce) cans minced clams
2 tablespoons lemon juice
1 yellow or white onion, cut into large pieces
1 green bell pepper, seeded and
        cut into large pieces
2 cups fresh parsley
4 garlic cloves
1 tablespoon oregano, dried
¼ pound (1 stick) butter
1 ¼ cup Italian-style bread crumbs
¼ cup Parmesan cheese
3 slices (3 ounces) American cheese, shredded
Paprika to taste

Preheat oven to 450 degrees.
In a medium saucepan or skillet over medium-low heat, simmer clams (with juice) and lemon juice for 15 minutes.
Using a blender or food processor, blend onion, pepper, parsley, and oregano until it reaches a thick soupy consistency.
In a large saucepan over medium heat, melt butter and add vegetable mixture and sauté for several minutes, stirring occasionally.
In a large mixing bowl, mix clams, vegetable mixture, and bread crumbs thoroughly.
Pour this mixture into an 8- or 9-inch pie plate.
Sprinkle with Parmesan cheese followed by American cheese.
Sprinkle with paprika.
Bake for 15–20 minutes.
Remove from heat and let cool slightly.
Serve with your favorite crackers.

3 eggs, well beaten
1 cup vegetable oil
2 cups sugar
2 cups zucchini, grated
2 teaspoons vanilla extract
3 cups all-purpose flour
1 teaspoon baking soda
1/2 teaspoon baking powder
3 teaspoons ground cinnamon

Preheat oven to 325 degrees. In a large mixing bowl, combine eggs, oil, sugar, zucchini, and vanilla.
Mix these ingredients well.
Fold in the flour, baking soda, baking powder, and cinnamon until well blended.
Pour mixture equally into two separate, greased 9-inch loaf pans.
Bake for 1 hour.
Remove bread from pans immediately and cool on wire racks.
Slice and serve.

Use Your Tool
Unless you're a Muscle Bear who wants a good upper-body workout from mixing bread dough, an electric mixer is worth the investment for this recipe alone. If you like to bake other breads, you have more reason to own one.

More Bearable Meal Suggestions
Coffee and/or your favorite brunch foods.

ZUCCHINI BREAD

Frankly, I've never been fond of zucchini as a vegetable. However, I love to try any variation of a good zucchini bread recipe. This is one of the most traditional and my favorite.

GLAZED NUTS AND BERRIES

Bears are especially known for their acute grazing skills. This recipe was inspired by nature and a natural love for "anytime snacking." This also makes a wonderful cocktail party snack.

4 tablespoons butter
6 tablespoons maple syrup
2 teaspoons salt
1 teaspoon ground cumin
1 teaspoon cayenne pepper
¼ teaspoon ground nutmeg
1 pound (3 cups) whole blanched almonds
1 cup dried cranberries
¼ teaspoon salt

Preheat oven to 325 degrees.
In a medium-sized saucepan, melt butter over medium heat.
Add syrup, salt, cumin, cayenne, and nutmeg; simmer for 1 minute.
Add almonds and toss several times over heat until well coated.
Place almonds in a single layer on a baking sheet lined with heavy foil.
Bake at 325 degrees for 10 minutes, then stir to turn nuts.
Bake an additional 10 minutes.
Stir in cranberries and bake an additional 2 minutes.
Remove from oven and transfer mixture to another baking sheet breaking up any pieces that are stuck together.
Let cool.

Use Your Tool
Double the recipe and use your favorite airtight container to keep this delicious snack around the den all week—if it lasts that long.

1 pound butter, softened (room temperature)
1 ½ cups sugar
2 teaspoons vanilla extract
3 ½ cups all-purpose flour
1 ½ cups potato chips, crushed
2 large eggs

Preheat oven to 350 degrees.
Blend butter, sugar, vanilla, and eggs.
Add the flour and potato chips and blend well.
Drop mixture by teaspoonfuls onto lightly greased
cookie sheets approximately 2 inches apart.
Bake for 12-15 minutes.
Remove from oven and cool—preferably on wire racks.

More Bearable Meal Suggestions
Serve with a large glass of whole milk—naturally.

POTATO
CHIP
COOKIES

*I consider this
to be the ultimate
Bear snack.
C'mon. Potato
chips turned
into cookies?
I rest my case.*

DEEP-FRIED TWINKIES

This is the mother of all Bear snacks! With every delectable bite, I weep with joy.

6 Twinkies
6 Popsicle sticks
4 cups vegetable oil
all-purpose flour for dusting

Batter

1 cup whole milk
$1/8$ teaspoon vanilla extract
2 tablespoons vinegar
1 tablespoon vegetable oil
1 cup all-purpose flour
1 teaspoon baking powder
$1/2$ teaspoon salt

Chill or freeze Twinkies for several hours (or overnight). When ready to fry, heat oil to 375 degrees in a deep fryer or on the stovetop using a large stockpot until oil is hot. To make batter, mix milk, vanilla, vinegar, and oil in a small bowl. In a separate large bowl, blend flour, baking powder, and salt. Whisk milk mixture into the dry ingredients until smooth. Refrigerate batter until oil heats completely.
Push stick into Twinkie lengthwise leaving 2–3 inches exposed to use as a handle; dust with flour; dip into batter, rotating to fully coat. Place carefully into hot oil. Twinkie will float so turn it to ensure even cooking. Cook for 3–4 minutes or until golden brown. Remove from oil and drain on paper towels. Remove stick and let cool for at least 5 minutes before serving.

Use Your Tool

Don't risk burning your fur or skin. Please use extreme caution when cooking with hot oil!

More Bearable Meal Suggestions

Serve with your favorite dessert sauces: butterscotch, strawberry, or my personal favorite, *Chocolate Gravy, Baby* (page 53).

1 (18-ounce) box Golden Grahams cereal
1 (15-ounce) box golden raisins
3 cups mixed nuts
12 ounces semisweet chocolate morsels
2 cups creamy peanut butter
1 teaspoon vanilla extract
1/4 pound (1 stick) butter
1 pound confectioners' sugar

In a saucepan over medium-low heat, carefully melt butter, chocolate, and peanut butter. Be careful not to boil.
Remove from heat, stir in vanilla, and let cool.
In a large mixing bowl, combine cereal, raisins, and nuts.
Pour butter sauce over cereal mix and blend until well-coated.
Pour mixture into a large sealable container or brown paper bag.
Gradually pour confectioners' sugar into container, close, and shake until thoroughly coated.
Serve or store in tightly covered container.

WHITE TRASH STASH

This is another snack that falls into the "crack snack" category. I like to keep a container of it near my bed for midnight snacking. Then there's that other container near my TV remote controls, and that stash in the bathroom sink cabinet.
Uh, I think I've said too much.

JERKY APPETIZER PIE

This is just the thing to begin a hearty dinner with a few friends. It is best served hot with an assortment of breads and crackers. I like to spoon it onto warm biscuit halves—woof!

1 (8-ounce) package cream cheese, softened
2 tablespoons whole milk
1 (2-ounce) jar dried beef, chopped
2 tablespoons minced onion
1 tablespoon each green and red bell pepper,
　　　finely chopped
1/2 cup sour cream
1/4 cup pecans, chopped

Preheat oven to 350 degrees.
In a mixing bowl, blend cream cheese and milk.
Stir in onion, dried beef, bell peppers, and sour cream.
Spread mixture into a pie pan; sprinkle top with nuts.
Bake for 15 minutes; serve hot.

2 (11-ounce) cans cheddar cheese soup
1/4 cup dry white wine
2 cups (8 ounces) Swiss cheese, shredded
1/2 teaspoon garlic powder
1 teaspoon freshly ground black pepper
1/4 cup green onions, finely chopped
1/4 cup red onion, finely chopped

In a large saucepan over medium heat, combine soup, wine, and cheese, stirring occasionally until cheese is melted.
Add garlic, pepper, and onions and stir until well-blended.
Pour into fondue pot or chafing dish to keep warm. If fondue becomes too thick, dilute with some extra wine.

Use Your Tool

Considering this is fondue, the best tools are the foods you choose to dip with. I recommend bite-sized pieces of bell peppers (any color), broccoli, cauliflower, celery, cherry tomatoes, stuffed olives, and my personal favorite, chunks of fresh French bread.

DOUBLE CHEESE FONDUE

Some cultural observers brand it as "trendy." As far as I'm concerned, fondue has never been nor will be out of fashion. Break the dinner monotony and surprise your friends or your Husbear with a fondue dinner.

BEEFY BROILER SNACKS

Toaster ovens and oven broilers can be a Bear's best friend in the kitchen. They are great for quick hot snacks like this one. This is especially convenient for serving at parties.

1 pound pork sausage
1 pound lean ground beef
1/4 teaspoon cayenne pepper
1 tablespoon Worcestershire sauce
2 teaspoons dried oregano
1 pound Velveeta
1 loaf sliced cocktail rye bread

In a large skillet over medium-high heat, brown sausage and ground beef until fully cooked. Sprinkle with pepper while cooking.
Drain fat from meat.
Stir in Worcestershire sauce and oregano.
Cut cheese into large cubes; stir into meat mixture until melted.
Heat broiler.
Spread meat/cheese mixture on rye slices and place onto a baking sheet.
Broil for approximately 1 minute or until bubbly.
Serve hot.

Use Your Tool
Always watch food carefully when using a broiler to prevent burning.

3 (6-inch) pita breads
3 tablespoons dried parsley flakes
2 green onions, finely chopped
1 teaspoon olive oil
1 teaspoon dried basil, crumbled
½ teaspoon dried rosemary, crushed
½ teaspoon minced garlic from jar
      or 1 medium garlic clove, minced
regular or butter-flavored cooking oil spray
2 tablespoons Parmesan cheese, grated

Preheat oven to 350 degrees.
Separate each pita into single layers.
In a mixing bowl, combine parsley, green onions, olive oil, basil, rosemary, and garlic; blend well.
Spread mixture evenly over pitas.
Spray tops with cooking oil spray; sprinkle with cheese.
Cut each pita into 6 wedges.
Place on an ungreased baking sheet; bake for 10-12 minutes or until crisp.

Use Your Tool
Any leftovers can be stored in an airtight container for up to one week.

More Bearable Meal Suggestions
This versatile snack is a great addition to meals such as *Vineyard Pasta* (page 116) or *Italian Sausage Casserole* (page 127).

ITALIAN PITA CRISPIES

Despite some critics, not all Bear comfort foods are slathered with mayonnaise and cheese (although I prefer that!). This is a delicious and heart-healthy snack that is quick and satisfying.

35

Section II

WOOFY BREAKFAST

CORN AND BACON MUFFINS

These muffins are so impressive to serve. They're the first things devoured when I host a brunch.

1 ¼ cups all-purpose flour
¾ cup yellow cornmeal
1 tablespoon baking powder
½ teaspoon salt
1 cup whole milk
¼ cup (½ stick) butter, melted
1 egg
1 cup cooked corn kernels
6 slices bacon, cooked and crumbled
½ teaspoon dried parsley flakes
½ cup green onions, finely chopped

Preheat oven to 425 degrees.
In a large mixing bowl, combine flour, cornmeal, baking powder, and salt.
In a separate bowl, whisk together milk, butter, and egg; then stir in corn, bacon, parsley, and green onions.
Pour into flour mixture; stir until just moistened.
The batter should be lumpy.
Spoon batter into 12 greased or paper-lined muffin cups until ¾ full.
Bake 20-25 minutes until tops are golden brown.

Use Your Tool
Another great reason to invest in muffin pans!

More Bearable Meal Suggestions
Make it easy on yourself—serve with scrambled eggs and hot coffee.

4 cups ripe bananas, mashed
1/3 cup lemon juice
2 Tablespoons brown sugar
1/8 teaspoon ground cinnamon
1/8 teaspoon ground nutmeg

Combine all ingredients in a blender and puree until smooth.
Pour mixture into a saucepan over high heat until it boils.
Turn down heat and simmer slowly for approximately 10 minutes, stirring occasionally until mixture thickens.

BANANA JAM

Crazy over bananas? This delicious jam is wonderfully versatile. Serve warm over pancakes or waffles, or cool and spread over toast. It makes a wonderful addition to any brunch table.

39

FALL HARVEST SCONES

Here is a toast to the Gentlemen Bears...the ones who still say "please" when asking for thirds and fourths, brush the crumbs out of their beards after every meal, and open your beer bottle before their own. To them, I tip my teacup and offer a hearty scone for teatime. Napkins on laps, please.

2 cups all-purpose flour
6 tablespoons cold butter, cut into pieces
3 tablespoons sugar
2 tablespoons baking powder
1/4 teaspoon salt
1/2 cup dried cranberries, chopped
1/2 cup dried apricots, chopped
2 large eggs
1/2 cup whole milk
3 tablespoons slivered almonds

Preheat oven to 400 degrees.
In a large mixing bowl, combine flour, butter, sugar, baking powder, and salt.
Using a pastry blender or an electric mixer, cut in butter until its texture turns into a coarse grain.
Stir in cranberries and apricots.
In a small bowl, beat eggs.
Remove and reserve 1 tablespoon.
Add milk to egg in bowl and beat to blend.
Add to flour mixture and stir until dry ingredients are moistened.
Coat a large baking sheet with cooking oil spray.
Drop 12 heaping tablespoons of dough (2 inches apart) on the prepared baking sheet.
Brush each with reserved egg, and sprinkle with almonds.
Bake 12-15 minutes or until golden brown.
Remove from oven and cool on a wire rack.
Serve warm or at room temperature.

Use Your Tool
An electric mixer is a must in order to create better dough.

1 ½ pounds (1 ½ tubes) uncooked pork sausage
½ cup red onion, chopped
½ teaspoon garlic powder
4 eggs, beaten
2 cups mozzarella cheese
¾ cup milk
1 (8-ounce) package refrigerated crescent
      roll dough

Preheat oven to 425 degrees.
In a large skillet over medium-high heat, add onion and
garlic to sausage and brown until fully cooked.
Remove from heat, drain oil, crumble, and set aside.
In large mixing bowl, combine sausage, eggs, cheese,
and milk.
Lightly grease a 9 x 13-inch baking pan.
Place crescent roll dough flat in bottom of pan.
Pour sausage mixture over dough.
Bake for 15 minutes until bubbly.

More Bearable Meal Suggestions
Preferably served in bed…

STAY FOR BREAKFAST

It is morning and you feel surprisingly happy. You see him lying there sound asleep and you still have the urge to impress him when he awakens. Here's what you do…

41

EASY BREAKFAST TORTILLA BAKE

A friend of mine once asked me what he could make for breakfast to satisfy his hungry Husbear. I suggested this and he has been serving it every Sunday morning for years. In fact, my friend's Husbear took him to Puerto Vallarta as a thank you. Life is good.

1 tube (1 pound) uncooked pork sausage
4 large eggs, lightly beaten
¼ cup green onions, chopped
½ cup fresh mushrooms, thinly sliced
      and chopped
2 tablespoons butter, melted
1 (8–10 ounce) package tortilla chips,
      coarsely crumbled
2 cups cheddar cheese, shredded
1 (16–ounce) jar mild salsa

Preheat oven to 350 degrees.
In a large skillet over medium-high heat, brown sausage until fully cooked, and drain oil.
In a medium-sized bowl, combine sausage, eggs, green onions, and mushrooms.
Pour melted butter into an 8 x 8-inch baking dish and add sausage mixture.
Top with crushed chips and sprinkle with cheese.
Bake for 10-12 minutes or until eggs are set and cheese is melted.
Remove from oven; top with salsa and serve.

Use Your Tool
Use your judgment when choosing an appropriately sized baking dish. If you use a 9 x 13-inch dish, slightly increase your ingredient amounts to compensate for the extra space.

1 tube (1 pound) uncooked pork sausage
2 (11-ounce) packages frozen au gratin potatoes,
    defrosted per package directions
1 ½ cups all-purpose baking mix
²/₃ cup whole milk
2 large eggs, lightly beaten
¼ teaspoon dried parsley flakes
¼ teaspoon freshly ground black pepper
1 cup sharp cheddar cheese, shredded
¼ cup green onion, finely chopped
¹/₈ cup vegetable oil

In a large skillet over medium-high heat, brown sausage
until fully cooked.
Remove from heat, drain oil, crumble and set aside.
In a large mixing bowl, combine baking mix, milk, eggs,
parsley, and pepper.
Add potatoes, crumbled sausage, cheese, and green
onions.
Stir until batter is well-blended.
In a large skillet over medium heat, coat the bottom of
skillet with oil.
When skillet is hot, pour ¼ cup of batter into skillet for
each pancake.
Cook until small bubbles appear on top on pancakes.
Turn over and continue to cook until undersides are
golden brown.

Use Your Tool
A large spatula is more convenient when flipping these
hefty pancakes.

More Bearable Meal Suggestions
Hot coffee and buttered toast.

SAUSAGE
POTATO
PANCAKES

*Considering my
French/Irish
lineage, it only
makes sense that
my two favorite
foods are cheese
and potatoes.
This explains my
absolute obsession
with au gratin
potatoes. I try to
incorporate them
into any meal
and here is one
example.*

CLASSIC BLUEBERRY MUFFINS

Very few breakfast foods can offer the comfort and emotional security that these beautiful gems provide. The aroma from the oven is enough to put a smile on my face. This recipe is very easy and should persuade you from using the boxed mixes. Their flavor can't compete with homemade.

2 cups all-purpose flour
1/2 cup sugar
1 tablespoon baking powder
1/2 teaspoon salt
1 egg, beaten
1 cup whole milk
1/4 cup butter, melted
1/2 cup walnuts, finely chopped
1 cup frozen blueberries, rinsed

Preheat oven to 375 degrees.
In a large mixing bowl, combine 1 3/4 cups of flour, sugar, baking powder, and salt.
Add egg, milk, and butter, stirring gently.
Batter should not be smooth.
Add nuts and blueberries and blend gently.
Spoon batter into paper-lined or greased muffin pans, filling each cup 2/3 full.
Bake for 25 minutes.

More Bearable Meal Suggestions
Enjoy with a tall glass of cold milk or serve with *Stay for Breakfast* casserole (page 41).

10 cups potatoes, peeled and cut into
       large chunks
1 ½ teaspoons salt
8 slices bacon
3 large yellow onions, sliced
Freshly ground black pepper to taste

Place potatoes in a large pot and fill with water until it
covers the potatoes by a half inch.
Add salt to water and stir until dissolved.
Bring water to a boil and cook until tender,
approximately 15–20 minutes.
Drain and set aside to cool.
In a large skillet over medium-high heat, cook bacon
until evenly brown.
Remove from skillet and place on folded paper towels
to drain.
Return skillet to heat with bacon grease still in it.
Add potatoes, onions, and crumbled bacon; sprinkle
with pepper and cook over medium-high heat—stirring
occasionally—for approximately 25 minutes until
potatoes brown.

More Bearable Meal Suggestions
Serve with eggs cooked any style!

BACON
HASH
BROWNS

*This recipe always
reminds me of my
first roommate
in college. It was
the only thing his
mother taught
him to make. He
was as dumb as a
box of rocks. He
couldn't tie his
shoelaces or use a
washing machine
correctly, but he
could make this
very well. And he
did—practically
every other day at
any time of the
day or night. I've
been a fan ever
since.*

45

A TRIBUTE
The Egg

I think Paul Simon said it best when he wrote, "The mother and child reunion is only a motion away."

I believe there is something instinctually necessary in the ritual of egg consumption. This is apparent because we've·cooked and eaten eggs for civilizations. We've farmed them, studied them, pasteurized them, and ultimately mastered their use in the kitchen.

We seem to know everything about them and yet, they still hold a certain awesome mystery and power. Their versatility in cooking is unmatched, and I will always be grateful that I'm not allergic to them.

Since I'm on the subject, here are some helpful hints regarding their use:

The time an egg needs to be in the boiling water to become a perfect boiled egg depends on the weight and size of the egg: the small egg will need to cook between 3 and 4 minutes, and the large egg will need to cook between 4 and 6 minutes.

Always store eggs in the refrigerator. Bacteria inside or on the outside of an egg can multiply much faster at room temperature than in your refrigerator. An egg stored under refrigeration for one week will be fresher than one stored at room temperature for just one day! The inside of an egg may be bacteria free, while due to its porous nature, the shell may have a high bacteria count.

CHOCOLATE CHIP WAFFLES

There's no better way to welcome the weekend (or nurse a hangover) than with a special breakfast. Treat yourself without working so hard at it! This recipe is made simpler thanks to your trusty microwave oven.

2 ¼ cups all-purpose flour
1 tablespoon baking powder
½ cup sugar
½ teaspoon salt
1 cup semisweet chocolate chips
¾ cup (1 ½ sticks) butter
1 ½ cups milk
3 eggs, beaten
1 tablespoon vanilla extract

In a large mixing bowl, combine flour, baking powder, sugar, and salt.
In a microwave-safe bowl, heat chocolate chips and butter on HIGH (100%) power for 1 minute. Remove and stir.
Continue to microwave for additional 20-second intervals until smooth.
Cool to room temperature.
Stir milk, eggs, and vanilla into dry ingredients and blend thoroughly.
Add chocolate mixture and stir. Batter should be thick.
Apply batter to heated waffle maker carefully following the manufacturer's instructions.

Use Your Tool
Yes, there is no way around it. You should use a waffle maker for this recipe.

More Bearable Meal Suggestions
Top with maple or chocolate syrup, whipped topping, and nuts. Serve with hot coffee, scrambled eggs, and bacon.

4 ripe bananas, peeled and sliced
8 slices hearty white or whole wheat
    sandwich bread
½ cup whole milk
2 large eggs
1 teaspoon vanilla extract
Pinch of ground cinnamon

Preheat oven to 350 degrees.
Place sliced bananas in a mixing bowl and mash them thoroughly.
Spread mixture onto 4 bread slices and place the remaining bread slices over each to make 4 sandwiches. Press each gently but firmly to ensure that bread slices adhere to the banana mixture.
In a large bowl, whisk together milk, eggs, vanilla, and cinnamon.
Place each sandwich in egg mixture and coat evenly. Let each side soak for 1 minute.
Spray large skillet with nonstick cooking oil spray (or melt a tablespoon of butter) and place over medium heat.
Add each sandwich, one at a time, to the skillet and cook each side approximately two minutes.
Transfer each sandwich to a baking sheet and bake all for 8-10 minutes.

Use Your Tool

A food processor, electric mixer, or a blender will blend the bananas much faster. Otherwise, mash with a fork the old-fashioned way.

More Bearable Meal Suggestions

Dust with powdered sugar and serve with maple syrup and *Chocolate Gravy, Baby* (page 53).

BAKED BANANA FRENCH TOAST

I love playing around with classic recipes. The warm, gooey bananas really take this up a notch. Who says you have to settle for regular French toast?

SWEET POTATO PANCAKES

This idea marries two of my favorite meals—breakfast and Thanksgiving dinner. Sweet potatoes have always been one of this Southern boy's favorite holiday side dishes. I guess I just couldn't get enough.

2 cups all-purpose flour
2 tablespoons sugar
2 teaspoons baking soda
1/2 teaspoon salt
1 teaspoon ground cinnamon
1/4 teaspoon ground nutmeg
1 1/2 cups whole milk
1 cup sweet potatoes, cooked, peeled, and mashed
1/2 cup currants
2 large eggs
1/2 cup (1 stick) butter, melted and cooled
1/2 teaspoon vanilla extract
Vegetable oil for griddle

In a large bowl, sift together flour, sugar, baking powder, salt, cinnamon, and nutmeg.

In a second large bowl, whisk together milk, sweet potatoes, currants, eggs, butter, and vanilla. Make a well in the center of the dry ingredients and add the sweet potato mixture.

Gently stir together with a large spoon until completely blended—being mindful not to overmix.

Place a large skillet or griddle over medium-high heat and lightly coat with oil.

Ladle batter onto the hot griddle using a 1/4-cup measure. Gently flip pancakes over once bubbles appear on the top of the batter and cook until both sides are lightly browned.

Use Your Tool

A large griddle is best for cooking more pancakes at one time!

More Bearable Meal Suggestions

Serve with maple syrup and top with miniature marshmallows or dollops of marshmallow cream.

50

4 eggs
1 cup all-purpose flour
2 cups whole milk
2 tablespoons cocoa powder

Filling
5 ripe bananas, mashed
1 tablespoon sugar
2 tablespoons brown sugar
2 tablespoons rum
1/2 cup pecans, finely chopped
6 tablespoons chocolate syrup

In a mixing bowl, combine filling ingredients and set aside.
In a large mixing bowl, whisk crepe ingredients.
Place a large skillet or griddle over medium-high heat and lightly coat with oil.
Pour batter onto griddle and spread thinly to the shape and size of a large pancake.
Cook for approximately 2 minutes and flip to cook the other side for 1 minute.
To assemble, add 1 tablespoon of filling to center of each crepe and spread evenly.
Fold or gently roll crepe and serve.

Use Your Tool
Remember, crepes are thinner than pancakes. The trick is in the batter pouring on the griddle—practice makes perfect!

More Bearable Meal Suggestions
Apply whipped topping and chocolate syrup to the top of crepes before serving! This also makes a perfect dessert.

COCOA CREPES

If my siblings and I were on our best behavior, my French mother would make crepes for Saturday breakfast, and enough to snack on for the rest of the day. For lunch, we would fill them with things like peanut butter and jelly or ham and cheese slices. I modified my mother's traditional recipe with cocoa to her dismay. Oh, Mon Dieu!

FRUIT COCKTAIL BREAD

And you thought fruit couldn't be hearty… Here is a way to enjoy eating fruit and still feel satisfied.

3 cups all-purpose flour
2/3 cup sugar
1 teaspoon baking soda
2 eggs, beaten
1 (8-ounce) container lemon yogurt
1/3 cup vegetable oil
1/2 teaspoon orange zest
1/2 teaspoon lemon zest
1 (15-ounce) can fruit cocktail, drained
2/3 cup walnuts, chopped

Preheat oven to 350 degrees.
In a large mixing bowl, combine flour, sugar, and baking soda thoroughly.
In a separate bowl, blend eggs, yogurt, oil, and zests.
Stir this mixture into the dry ingredients.
Add fruit cocktail and nuts.
Blend all ingredients gently—do not overmix!
Poor mixture into a greased 9 x 5-inch loaf pan and bake for 60-70 minutes.
Let cool in pan for 10 minutes.
Remove from pan and let cool on a wire rack prior to serving.

More Bearable Meal Suggestions
This is especially convenient as a last-minute gift when visiting friends or family. Trust me, it will impress.

½ cup butter
4 tablespoons unsweetened cocoa powder
⅓ cup all-purpose flour
⅛ teaspoon ground cinnamon
¾ cup sugar
2 cups whole milk

In a medium-sized skillet over medium heat, melt butter.
Add cocoa powder, flour, and cinnamon; stir until a thick paste is formed.
Stir in sugar and milk.
 Cook approximately 15 minutes, stirring constantly until thick.

More Bearable Meal Suggestions

As I previously mentioned, this is very versatile. I also like to pour it over desserts, and even over grilled meats like chicken or pork.

CHOCOLATE GRAVY, BABY

There are countless uses for this deliciously decadent gravy. It's all up to your imagination. I love to pour it over hot-buttered toast for breakfast. Yeah, baby!

53

HAM AND CHEESE MUFFINS

These delicious muffins are so easy to make and they're extremely versatile. I first began serving them for breakfast then began serving them during lunch and dinner meals. They also make a great snack and go very well with hearty soups.

2 cups self-rising flour
1/2 teaspoon baking soda
1 cup whole milk
1/2 cup mayonnaise
1/4 teaspoon dried parsley flakes
1/2 cup ham, finely chopped
1/2 cup extra-sharp cheddar cheese, shredded

Preheat oven to 425 degrees.
In a large mixing bowl, combine flour and baking soda.
Add the remaining ingredients and mix until blended thoroughly.
Spoon batter into 12 greased or paper-lined muffin cups and bake for 16-18 minutes.

More Bearable Meal Suggestions

For breakfast, serve with your favorite-styled eggs. For a quick lunch or dinner, serve with chili or *Winter Soup* (page 106).

3 tablespoons butter, divided
2 ½ tablespoons all-purpose flour
2 cups milk
½ teaspoon salt
¼ teaspoon freshly ground black pepper
½ cup cheddar cheese, shredded
½ cup American cheese, shredded
1 cup Canadian bacon, finely diced
1 cup fresh mushrooms, sliced
¼ cup green onion, chopped
⅓ cup yellow or white onion, chopped
1 dozen eggs, beaten
¼ cup melted butter
1 cup cornflakes cereal, crushed

To make cheese sauce, melt 2 tablespoons of butter in a medium-sized saucepan.
Blend in flour and cook for 1 minute.
Gradually stir in milk and cook, stirring occasionally until thickened.
Remove mixture from heat.
Stir in salt, pepper, and cheese and let stand.
In a large skillet over medium heat and using the remaining tablespoon of butter, brown Canadian bacon and mushrooms until golden brown.
Add onions and eggs and cook eggs until slightly ("soft") scrambled.
Stir cheese sauce into eggs and pour entire mixture into a greased 9 x 13-inch baking dish.
In a small bowl, combine ¼ cup melted butter and crushed cornflakes.
Sprinkle cornflakes over egg mixture.
Refrigerate overnight.
Bake uncovered at 350 degrees for 30-40 minutes.

BAKIN' & EGGS

This is a terrific breakfast casserole that is perfect to serve to overnight guests. It's better to prepare the night before so you don't have to work in the morning when you are at your grumpiest.
Don't let the homemade cheese sauce intimidate you—it's easy and makes this recipe ultimately much better tasting.

The Refrigerator

The refrigerator is one of the most important, and ironically, the least-appreciated appliances in your kitchen. That is why proper refrigerator care and maintenance is so important to a well-functioning kitchen. Remember, a refrigerator is designed to store food not create science fair projects.

To help inspire you, here are some easy tips.

Do's

- For best odor control, place a new open box of baking soda in your fridge every three months. Also, a solution of baking soda and warm water works best when cleaning shelves.

- Remember to vacuum behind the bottom grating at least twice a year.

- If your refrigerator is next to your stove, it may be working harder because of the heat emanating from the stove. Try placing a piece of plywood between the appliances to absorb the heat.

- To keep sticky syrup or honey containers away from your shelves, use plastic lids from coffee cans as coasters.

Don'ts

- Don't store perishables like milk in the fridge door. It is one of the warmest places in the icebox. Also, transfer milk from cardboard to glass or plastic containers. This will help deter bacteria breeding.

- Don't keep bread in the fridge. It maintains a longer shelf life in your pantry unless you freeze it.

- Don't keep fruits like fresh berries in their original plastic containers. Repackage them in a sealed plastic bag for longer life.

- Don't wash your produce before refrigerating. It will mold and rot faster.

1 (15-ounce) can sliced yellow cling peaches
    in heavy syrup
3 slices (approximately 3 ounces) cooked ham,
    thinly sliced and chopped
1/3 cup precooked imitation bacon bits (or two
    slices of real bacon, cooked and crumbled)
4 frozen waffles, thawed
1/4 cup pecans, chopped

Drain peaches reserving 2 tablespoon of syrup.
In a medium-sized skillet over medium-low heat,
combine peaches, ham, bacon, and reserved syrup.
Stir until heated thoroughly.
Prepare waffles according to package directions.
Spoon hot peach mixture over waffles and top with
maple syrup when serving.

Use Your Tool
No waffle maker necessary! I would recommend a
toaster.

More Bearable Meal Suggestions
Serve with your favorite scrambled eggs and sausage.

I-95
WAFFLES

*U.S. Interstate
95 is one of my
favorite roadways
in America. It
runs the entire east
coast from Miami,
Florida, to Bangor,
Maine. This recipe
was inspired by
many special
places found in
between—from
southern pecans
and Georgia
peaches to the
Virginia ham and
sweet maple syrup
of New England.*

OATMEAL SURPRISE

This ain't your papa Bear's oatmeal…why settle for ordinary homemade oatmeal when you can make it extraordinary?

3 ½ cups whole milk
¼ teaspoon salt
2 cups rolled oats
⅓ cup dried cranberries
⅓ cup raisins
⅓ cup maple syrup
⅓ cup wheat germ
½ cup pecans, chopped

In a medium saucepan, bring milk and salt to a gentle boil. Watch carefully while boiling.
Stir in oats, cranberries, raisins, and syrup.
Return to a boil then reduce to medium heat. (Cook "quick" oats for 1 minute; cook "old-fashioned" oats for 5 minutes.)
Cook until most of the milk has been absorbed, stirring occasionally.
Let stand to thicken.
Add wheat germ and pecans and blend carefully.
Serve into bowls and drizzle additional maple syrup on top.

58

2 (21-ounce) cans apple pie filling
1 ½ cups frozen cherries
¼ teaspoon ground cinnamon
1 (12-ounce) can refrigerated biscuits
1 tablespoon butter, melted
2 teaspoons sugar
½ cup pecans, finely chopped

Preheat oven to 400 degrees.
In a large mixing bowl, combine pie filling, cherries, and cinnamon.
Pour mixture into a 9 x 13-inch glass baking dish.
Bake for 20-25 minutes.
Remove baking dish from oven.
Separate biscuit dough and arrange on top of hot fruit.
Brush biscuits with melted butter and sprinkle with sugar and pecans.
Return dish to oven and bake for an additional 15-20 minutes.
Biscuits should be golden brown.

More Bearable Meal Suggestions
Serve with bacon-cheddar or Denver Omelets.

APPLE CHERRY PANDOWDY

The origin of the name "pandowdy" is unclear, although some seem to think it comes from this deep-dish dessert's dowdy (plain and old-fashioned) appearance. Since I'm notorious for breaking food traditions, I love serving this for breakfast instead of dessert— especially with cheesy, meaty omelets.

CARAMEL APPLE PANCAKES

Hey, it doesn't get much better than this. I rank this as one of the best Sunday morning breakfasts anyone can eat.

Pancakes

2 medium Granny Smith apples, unpeeled, cored, shredded
²/₃ cup buttermilk
2 large eggs
3 tablespoons butter, melted
3 tablespoons sugar
1 teaspoon vanilla extract
3 tablespoons sour cream
½ teaspoon baking soda
1 ¼ cups cake flour
¼ teaspoon salt
½ cup vegetable oil

Preheat oven to 200 degrees.
Shred apples using a food processor (ideally) or a manual grater and set aside.
Combine buttermilk, eggs, melted butter, sugar, and vanilla in food processor (ideally) or blender and pulse several times until slightly mixed. Add apples, sour cream, baking soda, cake flour, and salt and pulse processor several more times—do not overprocess.
In a large skillet over medium heat, coat the bottom of skillet with oil. When skillet is hot, pour ¼ cup of batter into skillet for each pancake. Cook until small bubbles appear on top of pancakes. Turn over and continue to cook until undersides are golden brown.
Store cooked pancakes in oven until ready to serve.

Caramel Apple Syrup

2 tablespoons butter
2 medium Granny Smith apples, cored and diced
1 teaspoon sugar
1 cup maple syrup

In a medium-sized skillet over medium-high heat, melt butter until hot. Add apples and sugar. Stir occasionally until apples caramelize, approximately 6-7 minutes. Add maple syrup and simmer for 1 minute. Pour over pancakes and serve.

1 (8-ounce) package pitted dates, chopped
¼ cup pecans, finely chopped
¼ cup walnuts, finely chopped
Zest of one large orange
2 cups plus 2 tablespoons all-purpose flour
1 teaspoon each baking powder,
        baking soda, salt
1 cup sugar
½ stick (¼ cup) butter, softened
1 egg
1 cup orange juice
3 (14 ½-ounce) soup cans

Preheat oven to 350 degrees.
In a small bowl, combine dates, nuts, orange zest, and 2
tablespoons flour and set aside.
In another bowl, blend 2 cups of flour, baking powder,
baking soda, and salt and set aside.
Using preferably an electric mixer on medium speed,
beat together sugar and butter until light and fluffy,
approximately 3-4 minutes; beat in the egg.
Add ½ cup of flour mixture to butter mixture; beat in
¼ cup orange juice.
Repeat, alternating flour and orange juice, beating well
after each addition. Stir in dates and nuts.
Spoon the batter into 3 greased and floured 14 ½-
ounce cans, filling each three-quarters full.
Place them on a baking sheet; bake until tops are
puffed and a toothpick inserted in the center comes out
clean—approximately 50 minutes.
Let cans cool on wire rack for 5 minutes. Remove bread
from cans and allow to cool completely on rack before
serving.

Use Your Tool
Like many bread/baking recipes, a batter is best created
using an electric mixer. Also, you can use a greased and
floured coffee can (or a 9 x 5-inch loaf pan) to turn this
into a loaf recipe!

OLD-FASHIONED SOUP CAN BREAD

*In my opinion,
this recipe
represents one of
the best examples
of American
cooking ingenuity.
Give this a try the
next time you're
about to throw
away a soup or
coffee can. You'll
be glad that you
did.*

SPAM IT, DAMN-IT!

This is for all of the cranky Bears who wake from their hibernations much too soon.

1 (12-ounce) can SPAM luncheon meat, chopped
2 large eggs, beaten
20 saltine crackers, crushed
2 cups whole milk
1 ¼ cups extra-sharp cheddar cheese, grated

Preheat oven to 350 degrees.
In a large mixing bowl, combine all ingredients with 1 cup of cheese thoroughly.
Pour into a 9 x 13-inch baking dish.
Sprinkle remaining ¼ cup of cheese on top, and bake uncovered for 1 hour.

More Bearable Meal Suggestions
This qualifies as a dinner entrée as well. If you serve for breakfast, include some hot-buttered toast and coffee on the side.

62

1 cup apple juice
1 frozen banana
3 heaping tablespoons rolled oats
½ teaspoon ground cinnamon
3 tablespoons maple syrup (or more to taste)

Add all ingredients into blender and blend until smooth.

Use Your Tool
A blender is a must!

More Bearable Meal Suggestions
If I've been a very good Bear, I add vanilla ice cream… yum.

BREAKFAST ON THE RUN

Drink your breakfast? Of course you can! This will give you a great energy boost in the morning, especially after a hard night.

Section III

MORE HEARTY SIDES

BACON & BUTTERMILK MASHED POTATOES

These are no ordinary potatoes. The bacon gives this dish its crunch while the buttermilk adds some tang. It's a winning combination!

6 bacon slices
4 pounds russet potatoes, peeled
    and quartered
1 cup buttermilk
2 tablespoons (1/4 stick) butter
2 tablespoons green onion, chopped
1/4 teaspoon each salt and fresh ground
    black pepper

In a large skillet over medium-low heat, cook bacon until crisp—approximately 7–8 minutes.
Drain on paper towels, crumble and set aside.
Reserve 2 tablespoons bacon fat.
In a large pot of boiling salted water, cook potatoes until tender—approximately 30 minutes.
Drain well.
Mash potatoes in same pot.
Add buttermilk, butter, reserved bacon fat, and mash again.
Add green onions, salt and pepper, and stir to blend.
Transfer to serving dish and serve.

Use Your Tool
If you don't own a potato masher, buy one. Also, an electric mixer can mash well, but only with the correct attachment.

More Bearable Meal Suggestions
Serve with *Barbeque Meat Loaf* (page 142) or *Oven-baked Pork Chops* (page 161).

½ pound penne pasta
1 lemon
1 tablespoon minced garlic
¼ cup fresh basil, chopped
1 (6-ounce) can tuna in oil, drained
½ teaspoon salt
¼ teaspoon freshly ground black pepper

Cook pasta as directed on package; drain well and cool slightly.

Grate enough lemon zest for 1 tablespoon and squeeze the lemon for 1 teaspoon of juice; add both to a large mixing bowl.

Add garlic, basil, tuna, salt and pepper and mix well. Add cooked pasta to tuna mixture and toss lightly.

Use Your Tool

I recommend that you cook the pasta until al dente (firm).

TUNA LEMON PASTA

I consider this a "side dish" because I love to offer it at potluck occasions. It's a great summer party favorite.

TRIPLE CHEESE MAC ATTACK

This is one of the most-revered comfort foods in American cooking. I recommend that you experiment and make this recipe your own.

2 cups cottage cheese
1 (8-ounce) cup sour cream
1 egg, lightly beaten
1 teaspoon salt
Garlic salt and freshly ground black pepper
    to taste
2 (8-ounce) cups extra-sharp cheddar cheese,
    shredded
½ cup Swiss cheese, shredded
1 (7-ounce) package elbow macaroni,
    cooked and drained

Preheat oven to 350 degrees.
In a large mixing bowl, combine cottage cheese, sour cream, egg, salt, garlic salt, and pepper.
Add cheddar and Swiss cheeses and blend well.
Cook macaroni as directed on package and drain well.
Add cooked macaroni to cheese mixture and blend until well-coated.
Add macaroni mixture to a greased 2-2 ½-quart baking dish and bake uncovered for 25-30 minutes.

More Bearable Meal Suggestions
This is a remarkably versatile side dish. It can accompany any meat dish and is especially good with seafood such as *Even I Can Fry Fish* (page 145) or *Golden Fish Sticks* (page 162).

1 cup uncooked instant white rice
1 cup water
1 tablespoon vegetable oil
¼ cup red onion, chopped
2 (16-ounce) cans baked beans
1 (4-ounce) can chopped green chiles
1 (1-ounce) package taco seasoning mix
½ cup sharp cheddar cheese, finely shredded

Cook rice as directed on package.
In a seperate medium-sized saucepan over medium-high heat, add oil and heat until hot.
Add onion, cook and stir for 1 minute.
Reduce heat to medium and add beans, chiles, and taco seasoning; mix well.
Cook until mixture begins to boil while stirring occasionally.
Add cooked rice and cheese to bean mixture and blend well.
Transfer to serving dish.

More Bearable Meal Suggestions
As an extra touch, offer a serving bowl of fresh sour cream to apply to the top of single servings.

TIJUANA BEANS AND RICE

The taco seasoning and baked beans help to make this dish fast, easy, and full of flavor. One bite and you'll be howling, "Si, si! papi Bear!"

69

FIESTA SOUP

Inspired by a trip to Cancun, this soup is guaranteed to warm you up on a cold winter day.

2 cups chunky salsa, medium or hot flavored
2 (15-ounce) cans black beans,
     drained and rinsed
2 cups frozen corn
2 cups deli-style turkey or chicken,
     thickly sliced and chopped
2 (15-ounce) cans chicken broth

In a large pot over medium-high heat, combine all ingredients and bring to a boil.
Reduce heat and simmer for 6-8 minutes.

Use Your Tool
Garnish with fresh sour cream and sprinkle with chopped cilantro.

More Bearable Meal Suggestions
Serve as a first course before serving grilled meats like *Marinated Flank Steak* (page 156).

5 cups water
1/2 cup sugar
6 or 7 tea bags

Heat water to boiling.
For simple syrup, pour 1 cup of water into glass measuring bowl; stir in sugar until it dissolves. Set aside.
Pour remaining water into teapot or additional glass measuring bowl with tea bags; steep 5 minutes. Remove the bags.
Stir syrup into the tea and pour tea into 4 ice-filled glasses.

SWEET ICED TEA

As a Southern child, I learned at a very early age why sweetened ice tea was considered the "mother's milk" of the South. I remember drinking more of it than water for most of my childhood. It was served at practically every meal, and I am surprised that I didn't pour it over my breakfast cereal. With that said, I am including it here and officially promoting it as a side dish. It just makes sense to me. There are many variations practiced by loyal sweetened-iced-tea drinkers. This one is tried and true.

CHICKEN AND CORN CHOWDER

Nothing impresses guests more than a homemade soup or chowder. They don't need to know how easy it was to make.

1 tablespoon vegetable oil
2 yellow or white onions, chopped
2 (14-ounce) cans chicken broth
2 ex-large russet potatoes, peeled and cut into
    ½-inch pieces
2 (10-ounce) packages frozen corn
2 cups cooked chicken breast meat, chopped
3 teaspoons dried thyme
2 cups half-and-half

In a large pot over medium heat, sauté onions in oil until they begin to soften, approximately 3-4 minutes.
Add broth and potatoes.
Cover pot and simmer for 5-6 minutes.
Add corn, chicken, and thyme and simmer, partially covered, for 10 minutes; stirring occasionally.
Add half-and-half and simmer an additional 4 minutes.

Use Your Tool
Fully cooked and cut chicken meat is a convenient option for this recipe and can be found in your grocery's meat section.

YAM-PPLE CASSEROLE

Maybe it's the Southern boy in me, but I love yams. And I make no apologies for it! They're sweet and starchy and full of flavor.

1 pound yams, peeled and sliced
2 large green apples, peeled, cored, and sliced
½ cup apple juice
½ cup apple sauce
2 tablespoons cornstarch
3 tablespoons water
⅓ cup sugar
½ teaspoon ground cinnamon
⅓ cup wheat germ

Preheat oven to 350 degrees.
Grease a 9 x 13-inch baking dish.
Alternately layer sliced yams and apples and set aside.
In a medium saucepan over medium heat, combine apple juice and apple sauce.
In a small bowl or cup, mix cornstarch and water.
When juice begins to boil, stir in cornstarch mixture.
Cook and stir until mixture has thickened.
Stir in sugar and cinnamon.
Spoon sauce over sliced yams and apples.
Sprinkle with wheat germ.
Bake for 25–30 minutes.

Use Your Tool
I know. You're probably thinking, "wheat germ?" Trust me. It makes a great crust when baked. You should try it on ice cream—it gives it a great crunch!

More Bearable Meal Suggestions
This goes well with roasted meats. Don't wait until Thanksgiving to enjoy it.

73

DADDY'S WALDORF RING

This is one recipe every civilized Cub should know and serve when his Bear daddy plays cards with his Bear buddies. I'm speaking of Saturday-afternoon bridge, not Friday night poker.

1/3 cup sugar
4 envelopes unflavored gelatin
5 cups white grape juice
3 medium celery stalks
1 pound seedless red grapes
3 medium Granny Smith apples,
      peeled and cored
3/4 cup mayonnaise
1/4 cup milk
1 cup walnuts, coarsely chopped

In a large saucepan over medium heat, combine sugar, gelatin, and 2 cups white grape juice; bring to a boil stirring constantly until gelatin is completely dissolved. Remove saucepan from heat and stir in remaining grape juice. Refrigerate for approximately 45 minutes or until mixture mounds when dropped from a spoon.
Meanwhile, chop celery and cut each grape in half.
When gelatin is ready as previously described, quickly shred apples with a grater or food processor (shredding any sooner will brown the apples).
Fold celery, grapes, and apples into thickened gelatin mixture; pour into 10-inch Bundt pan or 10-cup mold. Cover and refrigerate to set, approximately 3 hours. Unmold gelatin onto a chilled platter.
Serve with walnut dressing.

For walnut dressing: In a small bowl, mix mayonnaise and milk until thoroughly blended; stir in nuts.

Use Your Tool
There are many tricks to "unmold" gelatin molds. Try this: make certain that gelatin is completely firm -- it should not feel sticky on top and should not sag toward side if mold is tilted. If gelatin is firm, dip a small pointed knife in warm water and run tip of it around top edge of mold to loosen. Or moisten tips of fingers and gently pull gelatin from top edge of mold.

74

4 slices bacon
5 cups frozen corn, completely thawed
¼ cup butter
1 teaspoon salt
½ teaspoon freshly ground black pepper
2 large tomatoes, sliced

Preheat oven to 350 degrees.
In a large skillet over medium heat, cook bacon until crisp and brown.
Place on paper towels to drain and set aside.
Drain bacon grease from skillet.
In the same skillet over medium heat, melt butter; add corn and cook 5 minutes stirring constantly.
Stir in crumbled bacon, salt, and pepper and remove from heat.
Spread a layer of corn mixture in the bottom of a 2-quart casserole dish, then layer with sliced tomatoes.
Repeat layers twice ending with tomatoes on top.
Bake uncovered for 30 minutes.

CORN & TOMATO CASSEROLE

Get ready to enjoy one of the simplest and most versatile side dishes ever created. It complements so many meals, but is especially welcome at backyard barbeques.

KITCHEN TIPS

Cooking and Storing

Here are some extra helpful hints and tricks to make you more comfortable in your kitchen:

- To keep cleaned onions and radishes fresh longer, store them in a sealable plastic bag stuffed with several paper towels.

- To ripen tomatoes overnight, place them in a brown paper bag in a dark pantry.

- To allow more fridge space, stack food dishes using baking sheets as shelves.

- Add a few drops of lemon juice to simmering rice to keep grains separate.

- Pack nonfragile foods in sturdy sealable bags. Squeeze out the air and press food flat when sealing. Flat bags can be stacked easier.

- When boiling corn, add milk to sweeten the corn or sugar to soften. Salted water will toughen it.

- To drain cooked ground beef faster and easier, place it in a colander and shake gently over your kitchen sink.

- Soak onions or whole garlic in cool water for several minutes and their peels slip right off—with no onion "tears."

1 (15-ounce) can corn
1 (15-ounce) can cream-style corn
1 (8-ounce) container sour cream
1 cup melted butter
2 eggs, beaten
1 teaspoon dried parsley flakes
¼ teaspoon freshly ground black pepper
1 (8-ounce) package dry corn muffin mix

Preheat oven to 350 degrees.
In a large mixing bowl, combine all ingredients and stir until well blended.
Pour mixture into a 9 x 13-inch baking dish.
Bake for 35-45 minutes.

SOUTHERN-BAKED CORN

Yet another recipe from my Southern roots…You can't miss with this one. It goes well with any dinner and it's so easy to make!

HOT GRILLED STUFFED POTATOES

Impress your guests at your next outdoor party by showing off these beauties on the grill. Even your grilled meat will be jealous.

4 large baking potatoes
1 (28-ounce) can favorite style/brand
    baked beans
1 cup tomatoes, diced
3/4 cup green onions, chopped
1/2 tablespoon dried parsley flakes
1/8 teaspoon Tabasco sauce
1/8 teaspoon dried sage
Salt and freshly ground black pepper to taste
1 cup extra-sharp cheddar cheese

Wrap potatoes in heavy-duty foil and grill over medium heat for 1 hour, turning every 15 minutes. Allow to cool slightly.
Slice each warm potato in half lengthwise and scoop out most of potato, being careful not to break through the skin.
In a large mixing bowl, combine remaining ingredients except cheese.
Place mixture into potato skins and sprinkle with cheese.
Set potatoes on medium-heat grill for 10 minutes and serve.

Use Your Tool
To speed potato baking prior to final grilling, place washed and dried potatoes in microwave oven. Pierce each several times with a fork. Microwave on HIGH setting 12-15 minutes. Halfway during cooking time turn potatoes over.

More Bearable Meal Suggestions
Best served with grilled meats like *Marinated Flank Steak* (page 156) or *Texas Barbeque Shrimp Kabobs* (page 165).

The craving for comfort food continues to flourish within our popular culture as evidenced in the success of the decades-old, mom-and-pop lunch counters sprinkled throughout rural America as well as the newer urban retro diners.

Many of these places have a rich and dynamic culture and language of their own. For over a century, Americans have been served their favorite foods with waiter/waitress lingo that eventually worked itself into our standard language (i.e., BLT, "mayo," "moo juice," etc.).

If you are not already a regular patron of a local diner or lunch counter, become one. But before you do, you might want to brush up on the "lingo" before your next visit. The look of shock and delight on your server's face will be thanks enough. Try these on for size and have fun!

Eve with a Lid On—*Apple Pie*

Gentleman Will Take a Chance—*Corned Beef Hash*

Irish Turkey—*Corned Beef and Cabbage*

Bullets—*Baked Beans*

Draw One—*Coffee*

The Twins—*Salt and Pepper Shakers*

First Lady—*Spare Ribs*

In the Alley—*Serve as a Side Dish*

Wreck 'em—*Scrambled Eggs*

Put Out the Lights and Cry—*Liver and Onions*

Baby or Cow Juice—*Milk*

Frenchman's Delight—*Pea Soup*

Adam and Eve on a Raft—*Two Poached Eggs on Toast*

Bowl of Red—*Chili*

79

HUSH PUPPIES

*We southerners
are known for
the way we use
cornmeal—this
is my favorite
way. In fact,
these remind me
of all of the great
coastal seafood
restaurants
along the Florida
Panhandle where I
grew up.*

2 cups cornmeal
1 tablespoon baking powder
1 cup all-purpose flour
1 tablespoon salt
2 tablespoons sugar
1 ½ pounds yellow onions, minced
1 cup evaporated milk
3 large eggs
Water, if needed

In a large mixing bowl, combine all dry ingredients.
Add onion, then milk and eggs, stirring until
completely blended.
If batter is too stiff, hush puppies will be too dry.
Add very small amounts of water if necessary.
Batter should not be too "wet" either!
Drop tablespoons of batter into the very hot oil of a
deep fryer or deep stovetop boiler saucepan and cook
12-15 minutes or until golden brown.
Remove and drain on paper towels.

Use Your Tool

Always use extreme caution when cooking with hot oil.
A countertop deep fryer is recommended. Also, a food
processor makes mincing the onions a breeze!

More Bearable Meal Suggestions

A fish fry wouldn't be complete without these gems.

1 (16-ounce) package frozen mixed vegetables
2 tablespoons butter
2 tablespoons walnuts, chopped
6 bacon slices, cooked and chopped
3 tablespoons Parmesan cheese, grated
1/4 teaspoon freshly ground black pepper

Cook vegetables by using the stovetop or microwave instructions on package; drain.
Place butter and nuts in a small microwaveable bowl.
Microwave on HIGH setting for 1 minute.
Remove from oven and stir.
Return to microwave and heat on HIGH setting for an additional minute.
Add vegetables and butter mixture to a serving dish.
Add bacon, cheese, and pepper.
Toss lightly and serve.

Use Your Tool
This is a great microwave recipe that can save you time when preparing a full-course meal.

More Bearable Meal Suggestions
Serve with *Maple Glazed Ham* (page 143) or *Easy-Bake Swiss Steak* (page 144).

NUTS ABOUT VEGGIES

Of course, Bears eat vegetables…as long as there's bacon in it!

SHOCKING PINK CRANBERRY RELISH

Are you dreading going home for the holidays? Shake things up a bit with this dramatic holiday dinner accessory. It's guaranteed to turn heads. C'mon, do you think they expected you to bring something boring to the table?

2 cups raw cranberries
1 small white onion, quartered
1/4 cup sugar
1/4 cup brown sugar
3/4 cup sour cream
2 tablespoons prepared white horseradish

In a food processor, grind onion and cranberries into a minced texture; pour into sealable plastic container or food storage bag.
Add remaining ingredients and mix well.
Refrigerate and serve chilled.

Use Your Tool

A food processor gives the best textural results for this relish. This can be frozen as well. Remember to move container from freezer to refrigerator at least 1 hour prior to serving.

1 (28-ounce) can favorite flavor/brand
    baked beans
8 large Vidalia or yellow onions
1/4 cup brown sugar
1/2 cup favorite flavor/brand barbeque sauce
2 strips precooked bacon, chopped
1 teaspoon dried parsley flakes
4 tablespoons butter, cut into 8 pieces
Freshly ground black pepper to taste

To prepare onions, cut stems (avoid cutting too deep on one side to act as a base) and remove peel.
Carefully hollow out onions while leaving the base intact.
Finely chop onion pieces removed from shells.
Combine chopped onion, beans, brown sugar, bacon, parsley, and barbeque sauce; fill each onion bowl with bean mixture.
Sprinkle each with pepper and brush with melted butter.
Arrange onions on grill away from fire or most intense heat.
Grill 45-60 minutes until onions are tender and golden brown.

Use Your Tool
Carving out onions isn't difficult—just don't rush.

More Bearable Meal Suggestions
Serve with any grilled meats and a side of delicious homemade potato salad.

BARBEQUED ONION BOWLS

I love baked beans—love 'em— and I'm always looking for new ways to enjoy them. This is a great way to "jazz up" a routine barbeque.

83

SAGE DRESSING

As in Bear Cookin', *I felt it appropriate to include another of my many dressing/ stuffing recipes. It's my favorite holiday meal side dish. This is easy enough to enjoy making year-round.*

1 stick (½ cup) butter
2 celery stalks, chopped
1 medium yellow onion, chopped
1 loaf (1 ½ pounds) stale white or
    whole-wheat sandwich bread, cubed
1 ½ teaspoons minced fresh sage
1 ½ teaspoons poultry seasoning
1 teaspoon dried parsley flakes
1 (14-ounce) can chicken broth
1 teaspoon salt
Freshly ground black pepper to taste

Preheat oven to 350 degrees.
In a medium-sized skillet over medium heat, melt butter and add celery and onion.
Cook until tender, approximately 10 minutes.
In a large mixing bowl, combine cubed bread, sage, poultry seasoning, and parsley.
Add vegetable mixture and blend carefully to avoid tearing bread cubes.
Add chicken broth, ¼ cup at a time, until dressing is moist but not wet.
Season with salt and pepper.
Place dressing in a greased 9 x 13-inch baking dish; cook 20 minutes.

More Bearable Meal Suggestions
Serve with a holiday turkey, ham, or a premade rotisserie chicken; add a side of *Southern-Baked Corn* (page 77).

1 (5-ounce) package dry scalloped potato mix
1 (10-ounce) package frozen mixed vegetables
½ cup white onion, thinly sliced
¼ teaspoon celery seed
2 ½ cups boiling water
2 teaspoons butter
½ cup (2-ounce) sharp cheddar cheese,
    shredded

Preheat oven to 400 degrees.
Place dry potatoes in a 2-quart casserole dish.
Sprinkle the envelope of sauce mix from package over
the potatoes.
Top with frozen vegetables, sliced onion, and celery
seed.
Pour boiling water over all and stir well until combined.
Dot with butter, cover and bake 35-40 minutes.
Uncover, sprinkle with cheese, and return to oven for 3
minutes until cheese melts.

More Bearable Meal Suggestions
This is perfect for potlucks because it's so quick and
easy!

POTATO VEGETABLE SCALLOP

*Boxed scalloped
potato mix can
be very handy for
weekday meals.
Here's a way to
dress it up—get
creative!*

A TRIBUTE
The Potato

Dear Spud:

I just wanted let you know how much I adore you, from your spotted brown complexion to your rich, starchy, carbohydrate-filled soul. I've loved you all of my life. I think you know that already, don't you?

We met as innocents, remember? You were mashed so light and fluffy, resting so perfectly on my Curious George lunch plate, like a beautiful white cloud covered in peas and chopped carrots. I was so eager and hungry seated in my high chair, shamelessly attacking you with my bare hands. You were so hot to the touch—that only made you seem more forbidden to me. Your smooth buttery flavor teased my taste buds and gently caressed my throat. I will never forget that meal. I had never felt so full and satisfied. Even then, I knew we were destined to dine together for the rest of my life.

Now you've made it impossible for me to enjoy any of the other starches. I can only carbo-load with you, my dear sweet spud. I love you with and on everything I eat. I want you with me during every meal whether you're mashed, sliced, diced, boiled, fried, or twice baked.

I don't care how you look, damn it, I just want you.

I am writing this to thank you, and to let you know that we will soon be together. I just bought a few acres in Idaho and I want you to live with me.

Please say yes, my darling. I love you!

Most Sincerely,

PJ

TAILGATER SOUP

*Inspired by all
the Bears I know
who are devoted
to their favorite
football teams.
This is another
exceptional way
to keep beer in
your diet.*

2 tablespoons butter
10 packaged baby carrots
5 red potatoes, peeled and cut into large pieces
3 ribs celery, chopped
1 yellow onion, sliced
1/2 teaspoon salt
Freshly ground black pepper and
        red pepper flakes to taste
1 (12-ounce) bottle beer
2 (14-ounce) cans chicken broth
2 1/2 cups sharp cheddar cheese, grated

In a large saucepan over medium heat, melt butter.
Add carrots, potatoes, celery, and onion.
Cook, stirring, approximately 10 minutes until lightly
browned.
Add salt, pepper, and red pepper flakes to taste.
Add beer and heat to a boil.
Add chicken broth and heat to a boil.
Continue to cook approximately 12 minutes until
carrots and potatoes are tender.
Pour soup, in batches, into a food processor or blender
and puree.
Return to saucepan and stir in cheese.
Heat approximately 2-4 minutes until cheese is melted,
stirring occasionally.
Taste for seasoning and serve.

More Bearable Meal Suggestions
More beer, of course, and perhaps some warm *Cheesy
Cheddar Biscuits* (page 89).

2 cups all-purpose baking mix (Bisquick)
½ cup cold water
1 cup extra-sharp cheddar cheese, grated
¼ cup butter
1 teaspoon dried parsley flakes
½ teaspoon garlic powder
½ teaspoon Italian seasoning

Preheat oven to 450 degrees.
In a large mixing bowl, combine baking mix, cold water, and cheese.
Using a rolling pin or side of a large drinking glass, roll out biscuit dough on a floured surface.
Dough should be approximately 1-inch thick.
Using a cutter or top of a drinking glass, cut biscuits and place them on an ungreased baking sheet.
In a separate bowl, melt butter and seasonings; brush onto top of biscuits.
Bake for 8-10 minutes.

Use Your Tool
Dough is best mixed using a heavy-duty electric mixer.

More Bearable Meal Suggestions
Let's face it. One bite of these and you'll be ready to serve them with anything!

CHEESY CHEDDAR BISCUITS

These are inspired by a certain incredible biscuit served at a certain national seafood chain restaurant—the one with the "sunburned" lobster on the marquee. They claim to make theirs with a top-secret ingredient, but this comes awful close.

PUMPKIN CHEESE BAKE

If you love au gratin potatoes, you'll want to try this! It is my favorite way to eat pumpkin—with the exception of pumpkin pie.

1 (15-ounce) can pumpkin
2 whole eggs
2 egg yolks
1/2 cup Swiss cheese, grated
1/2 cup mild cheddar cheese, grated
1/2 cup whipping cream
1 tablespoon brown sugar
1/2 teaspoon salt
1/2 teaspoon ground nutmeg
Freshly ground black pepper to taste
2 tablespoons Parmesan cheese, grated

Preheat oven to 375 degrees.
In a food processor (ideally) or a blender, combine all ingredients expect Parmesan cheese.
Pulse to mix without overblending.
Pour batter into a greased 9 x 13-inch baking dish; sprinkle with Parmesan cheese.
Bake uncovered for 20 minutes or until golden brown.

Use Your Tool
A food processor offers the best results for mixing your batter.

More Bearable Meal Suggestions
This is great served hot or cold. I love eating it cold in a sandwich with cold cuts.

½ cup slivered almonds
½ cup vegetable oil
¼ cup white-wine vinegar
1 teaspoon vanilla extract
½ teaspoon salt
½ teaspoon freshly ground black pepper
1 pound carrots, grated
1 cup golden seedless raisins

Heat oven to 350 degrees.
Place almonds on a baking sheet in a single layer.
Toast until lightly brown for approximately 5 minutes.
Stir once for even coloring. Set aside.
In a large serving bowl, whisk together the oil, vinegar, and vanilla; add salt and pepper.
Add carrots, raisins, and toasted almonds and toss with the dressing.
Serve at room temperature or slightly chilled.

Use Your Tool
For best results when blending oil and vinegar, use a whisk instead of another utensil.

More Bearable Meal Suggestions
Best served with grilled chicken, pork, or fish.

SUMMERTIME CARROT SALAD

I love how the vanilla enhances the natural sweetness of the carrots. For me, these are the flavors of summer.

EASY SOUTHERN-STYLE GREENS

This is a variation of my family's recipe. The smoked ham and onions give these greens the right flavor. Damn, they're good!

1 ½ pounds Swiss chard, washed, dried
3 tablespoons vegetable oil
1 ½ cups yellow or white onion, chopped
⅓ pound thickly sliced, smoked deli ham, chopped
1 pound mustard greens, washed, trimmed, and coarsely chopped
1 cup chicken broth
2 cups frozen corn, thawed
Salt, freshly ground black pepper, lemon juice to taste
Cider vinegar, hot pepper sauce

Separate chard ribs from their leaves.
Chop leaves and stems coarsely.
In a large skillet over medium-low heat, add oil and onion.
Cook onion for approximately 3 minutes until softened.
Add ham, chard leaves and ribs, mustard greens, and broth.
Heat to a simmer, then cover.
Simmer for 10 minutes.
Add corn to skillet and cook uncovered approximately 5 minutes or until corn and greens are tender.
Stir in salt, pepper, and a few drops of lemon juice to taste.
Serve cider vinegar and hot sauce on the side.

More Bearable Meal Suggestions
Another great side dish to serve with barbequed meats or fried fish and a side of French fries or *Hush Puppies* (page 80).

92

YOUR OWN BARBEQUE SAUCE

2 cloves garlic, minced
1 large yellow or white onion, minced
1 cup vanilla-flavored cola
²/₃ cup ketchup
3 teaspoons brown sugar
2 teaspoons Worcestershire sauce
1 teaspoon hot pepper sauce
¹/₂ teaspoon ground mustard
¹/₄ teaspoon ground cloves

In a large saucepan over high heat, combine all ingredients and heat to a boil.
Reduce heat to simmer and cook, covered, for 1 hour.

Use Your Tool
A food processor is great for mincing, and a coffee bean grinder makes grinding whole cloves much easier.

More Bearable Meal Suggestions
Smother this all over your favorite meat. Enough said.

Oh, sure. You could buy a bottle of this at any supermarket. You may already have a favorite brand. But where is the fun in that? This is one of those recipes that you can play around with. Guests are always impressed when they discover that it's homemade!

93

PORKY'S CORN BREAD

My buddy, Porky, is a big Southern Bear and an ex-Marine who loves to cook with ham—hence his nickname (but that is questionable). This is one of his signature recipes. His favorite expression to me is, "Where you been, boy? We need to feed you some 'cone' bread!" Yes, sir!

1 (15-ounce) can creamed corn
3 eggs, beaten
1/2 pound cooked ham, cubed
1 cup all-purpose flour
1/2 cup cornmeal
3 green onions, chopped
1/4 cup butter, melted
1/4 cup whole milk
2 tablespoons sugar
2 teaspoons baking powder
1 cup extra-sharp cheddar cheese

Preheat oven to 350 degrees.
In a large mixing bowl, combine all ingredients—except the cheddar cheese.
Stir until just combined.
Pour into a greased 8-inch square baking dish.
Bake for 40 minutes until golden.
Sprinkle cheese on top; return to oven for 1-2 minutes.
Let stand 5 minutes before serving.

More Bearable Meal Suggestions

Porky likes to serve this with fried chicken, *My Favorite Deviled Eggs* (page 95), and *Potato Salad from Hell* (page 99).

8 hard-boiled eggs
3 tablespoons mayonnaise
1 tablespoon prepared mustard
1/2 teaspoon Worcestershire sauce
2 tablespoons minced sweet pickles
2 tablespoons sweet pickle juice
1 tablespoon cider vinegar
Salt and freshly ground black pepper to taste
Paprika to garnish

After eggs have boiled and cooled completely, peel the eggs and carefully cut each lengthwise, then remove yolks.
In a medium-sized mixing bowl, mash yolks.
Add remaining ingredients and mix until smooth.
Spoon mixture into each egg white cavity and sprinkle each lightly with paprika.
Chill before serving.

Use Your Tool
Here's a hard-boiled egg trick: When eggs have cooked properly, remove them and submerge them into a container of ice-cold water for several minutes. This helps make peeling easier!

MY FAVORITE DEVILED EGGS

This is another classic and one of my all-time favorite side dishes. I always notice that these are the first to go at a gathering.

TANGY COLESLAW

So many people associate coleslaw with summer foods and events. I like to eat this year-round. It's so easy.

1 ½ cups mayonnaise
½ cup white vinegar
⅓ cup sugar
1 tablespoon celery seeds
½ teaspoon lemon juice
1 teaspoon salt
Freshly ground black pepper to taste
1 head green cabbage, finely grated
2 carrots, finely grated
⅓ cup extra-sharp cheddar cheese,
        coarsely grated

In a large mixing and/or serving bowl, blend mayonnaise, vinegar, sugar, celery seed, lemon juice, salt, and pepper.
Add the cabbage and carrots and toss well. Refrigerate for at least 1 hour.
Before serving, add the cheese and toss lightly.

More Bearable Meal Suggestions
Serve with burgers and dogs from the grill with a side of *Barbequed Onion Bowls* (page 83).

96

1 pound penne pasta
3 cups your favorite pasta sauce
1 (15-ounce) container ricotta cheese
½ pound mozzarella cheese, shredded

Preheat oven to 450 degrees.
In a large pot, cook pasta al dente according to package directions.
Drain pasta and return to pot.
Add sauce and ricotta cheese and mix well.
Pour mixture into a 9 x 13-inch baking dish and top with mozzarella cheese.
Bake 20-30 minutes or until cheese is lightly brown.

ANYTHING GOES PASTA BAKE

Oh, the pasta-bilities…this extremely versatile recipe can be quite a chameleon. Sometimes I use penne or rigatoni or fusilli pasta. Sometimes I turn it into an entrée by adding meat or vegetables. You'll never be bored with this one.

QUICK MUSHROOM RISOTTO

It may not be a true risotto, but it's a fast and very delicious way to dress up boring instant rice. Don't thank me, man, thank the 'shrooms.

2 tablespoons butter
1 (4-ounce) jar sliced mushrooms, drained
2 cups uncooked instant white rice
1 teaspoon garlic powder
½ teaspoon dried parsley flakes
¼ teaspoon freshly ground black pepper
2 (14-ounce) cans chicken broth
⅓ cup half-and-half
⅓ cup Parmesan cheese, grated

In a large skillet over medium-high heat, melt butter; add mushrooms and cook, stirring occasionally, for 3-4 minutes.
Add rice, garlic powder, parsley, and pepper; cook 2 additional minutes.
Add 1 can of broth and cook 4 additional minutes, stirring constantly.
Gradually stir in remaining can of broth.
Cook an additional 7 minutes, stirring frequently.
Add half-and-half and cheese and remove from heat.
Blend thoroughly before serving.

More Bearable Meal Suggestions
Serve with *Oven-Baked Pork Chops* (page 161) or *Salmon Cracker Cakes* (page 170).

1 pound (4 medium) potatoes,
    cut into 1-inch cubes
3 tablespoons vegetable oil
3 teaspoons lime juice
2 tablespoons bottled jalapeño sauce (mild)
2 teaspoons chili powder
½ teaspoon salt
¼ teaspoon freshly ground black pepper
1 (15-ounce) can black beans,
    rinsed and drained
1 (7-ounce) can corn, drained
1 cup fresh tomatoes, diced
½ cup green onions, chopped

In a medium or large stockpot over medium-high heat, cook potatoes in boiling water for approximately 10-12 minutes or until just tender.
Drain and cool.
Meanwhile, in a large mixing bowl, whisk together oil, lime juice, jalapeño sauce, chili powder, salt, and pepper. Add potatoes and remaining ingredients and toss gently but thoroughly.

Use Your Tool
Skins on the potato pieces may be left on, but that is optional. Also, be careful not to overcook them so they remain firm when cooled and tossed.

More Bearable Meal Suggestions
Great for outdoor summertime meals! Serve with *Barbequed Onion Bowls* (page 83) and *Marinated Flank Steak* (page 156).

POTATO SALAD FROM HELL

Do you occasionally like a little fire in your meal? This has just enough kick to make you say a prayer or two before you leave the table. Do I hear an "Amen," brothers?

Section IV

COME-AND-GET-IT ENTRÉES

CITRUS HAM CASSEROLE

Growing up in the Sunshine State meant that an orange was likely to be incorporated into every meal. It's no surprise, then, that I love oranges. This recipe is proof.

1 large ham slice, thickly cut
6 sweet potatoes, peeled and cooked
¼ teaspoon salt
½ cup brown sugar
1 (6-ounce) can frozen orange juice
   concentrate, thawed, undiluted
3 tablespoons butter

Preheat oven to 350 degrees.
Peel and cut sweet potatoes into large chunks, place into a microwave-safe bowl.
Microwave potatoes on HIGH for 6-8 minutes.
Place ham slice and cooked potatoes into a 9 x 13-inch baking dish; sprinkle with salt followed by brown sugar.
Pour orange juice concentrate over ham and potatoes.
Dot with butter and bake uncovered for 40 minutes.

Use Your Tool
Ask your grocer's meat counterperson to slice the ham at least one inch thick. Also, the microwave should not cook the potatoes completely. That will be done in the oven.

More Bearable Meal Suggestions
Serve with *Cheesy Cheddar Biscuits* (page 89) and *Easy Southern-style Greens* (page 92).

102

4 cups (1 pound) cooked chicken breast meat,
    cut into 1-inch pieces
1 (16-ounce) bottle taco sauce
2 (4-ounce) cans diced green chilies
1 cup frozen corn
3/4 cup instant white rice
1/2 cup water
1 (2-ounce) can sliced ripe olives, drained
1/3 cup green onions, finely chopped
12 taco shells, crumbled
1 cup (8-ounces) extra-sharp cheddar cheese,
    shredded
1 cup (8-ounces) Monterey Jack cheese,
    shredded

Preheat oven to 375 degrees.
In a large mixing bowl, combine chicken, taco sauce,
chilies, corn, rice, water, olives, and green onions.
Spoon mixture into a 9 x 13-inch baking dish.
In a separate bowl, combine crumbled taco shells and
cheeses; sprinkle on top of chicken mixture.
Bake for 40-45 minutes or until top is golden brown.

Use Your Tool
Try replacing half of the taco shells with some lime-
flavored tortilla chips!

TACO CHICKEN CASSEROLE

*This is a great
alternative to
traditional tacos.
It was inspired
after the many
frustrating times
my tacos crumbled
in my hands. It
made me loco.*

SUPER BOWL STEAK AND BEANS

*Let's talk football.
Better yet, let's talk
players. Let's see,
you've got your
tight ends, your
wide receivers…
Did I forget
anyone? Now
let's talk food.
For Super Bowl
viewing, nothing
is more satisfying
than this.*

1 pound flank steak or skirt steak
    rubbed with olive oil
Salt and freshly ground black pepper
½ cup hickory-smoked barbeque sauce or
    Your Own Barbeque Sauce (page 93)

Bean Mixture

2 tablespoons olive oil
6 cloves minced garlic
1 teaspoon jalapeño peppers, minced
1 cup white or yellow onion, chopped
1 cup sliced mushrooms
½ cup red bell peppers, chopped
1 teaspoon dried parsley flakes
1 tablespoon chili powder
1 tablespoon brown sugar
½ cup ketchup
1 (15-ounce) can pinto beans, drained
1 (15-ounce) can chili beans, drained
½ cup green onion, chopped
½ cup extra-sharp cheddar cheese

Season oil-rubbed steak with salt and pepper.
Place on a preheated medium-high grill or under a broiler
for approximately 4 minutes on each side, or until desired
level of doneness. During the last few minutes of cooking
time, brush with barbeque sauce.
Remove from heat, cut meat into 1-inch chunks; set aside.
In a large saucepan over medium-high heat, cook garlic
in oil for 30 seconds. Add jalapeño peppers, onions,
mushrooms, red pepper, parsley, chili powder, and brown
sugar. Combine and cook for 2 minutes.
Add ketchup, pinto and chili beans, and steak pieces.
Combine and cook for 3 minutes. Place bean mixture in
serving bowls and top with green onion and cheese.

6 bacon slices
6 tablespoons butter
1 cup mushrooms, sliced
1/3 cup yellow or white onion, chopped
6 tablespoons all-purpose flour
1/4 teaspoon freshly ground black pepper
1 1/2 cups chicken broth
1 1/2 cups half-and-half
3 tablespoons dry sherry
8 ounces spaghetti
2 cups cooked turkey
1/3 cup Parmesan cheese

In a large saucepan, cook spaghetti in boiling water until al dente following the package directions.
Meanwhile, cook bacon, covered, in microwave on HIGH for 6-7 minutes.
Remove, crumble, and set aside.
Add butter, mushrooms, and onions to a large microwave-safe dish and cook on HIGH for 4-5 minutes.
Stir in flour and pepper; cook on HIGH for 30 seconds.
Stir in broth and half-and-half; cook on HIGH for an additional 9-10 minutes, stirring once after 5 minutes.
Blend in sherry and add spaghetti, turkey, bacon, and cheese.
Mix well and cook, covered, on HIGH for 14 minutes, stirring once.

More Bearable Meal Suggestions
As an obvious suggestion, serve with the rest of your holiday leftovers!

HOLIDAY HANGOVER TETRAZZINI

Decided to cook that elaborate holiday meal for your loved ones? And now you're exhausted? This is a great way to finish your leftover turkey immediately after the holidays. The microwave makes this easier to prepare when you have the least amount of strength to cook!

105

WINTER SOUP

It's so cold outside. You just want to hibernate in your den under warm, toasty blankets. But you're hungry, right? I know something perfect to make and take back to bed. This is so incredibly comforting that it should be shared.

1 (6-ounce) box long-grain and wild rice
1 tablespoon vegetable oil
2 (8-ounce) boneless, skinless chicken breasts, chopped
2 cups mushrooms, sliced
1 small red onion, chopped
2 cloves garlic, finely chopped
1 (14-ounce) can chicken broth
1/2 teaspoon dried tarragon, crushed
1/4 teaspoon dried thyme, crushed
1/8 teaspoon nutmeg
1 (12-ounce) can evaporated milk
2 tablespoons cornstarch
2 tablespoons dry white wine

In a saucepan, cook rice mix according to package instructions.
In a large saucepan over medium-high heat, add chopped chicken, mushrooms, onion, and garlic; cook, stirring occasionally until vegetables are tender and chicken is no longer pink.
Add rice, broth, tarragon, thyme, and nutmeg; bring to a boil over medium-high heat.
In a small bowl, combine a small amount of the evaporated milk and cornstarch; stir until smooth.
Add this mixture to saucepan and stir.
Add remaining milk and wine.
Cook until soup is thickened, stirring occasionally.
Season each serving with freshly ground black pepper.

More Bearable Meal Suggestions
Serve with *Cheesy Cheddar Biscuits* (page 89) or *Porky's Corn Bread* (page 94).

TARZAN SKILLET

Break out your spice rack because this dish is a flavor safari! It's so simple that even an ape-man could make it.

1 ½ pounds lean ground beef
1 small yellow or white onion, sliced
1 cup uncooked white rice
2 ½ cups water
2 teaspoons chicken bouillon
¼ cup green onion, finely chopped
1 teaspoon curry powder
½ teaspoon salt
¼ teaspoon ginger
¼ teaspoon cinnamon
3 tablespoons chunky peanut butter
2 tablespoons honey
½ cup seedless raisins

In a large skillet over medium-high heat, cook meat and onion until onion is tender.
Drain off fat.
Stir in remaining ingredients.
Heat mixture to boiling.
Reduce heat; cover and simmer, stirring occasionally, approximately 30 minutes or until rice is tender.

Use Your Tool
A small amount of water may be added if necessary.

107

20-MINUTE SHEPHERD'S PIE

Traditional Shepherd's pie is usually more elaborate and the baking and prep is usually much longer. By using instant potatoes, canned veggies, and one large skillet, this version is a snap without sacrificing great taste!

1 pound lean ground beef
Salt and freshly ground black pepper to season
1 (8-ounce) can sweet corn
1 (15-ounce) can sliced carrots
1 (15-ounce) can diced tomatoes
½ teaspoon minced garlic
½ teaspoon onion powder
4 servings instant mashed potatoes
½ cup sharp cheddar cheese, shredded

In a large skillet over medium-high heat, brown ground beef.
Sprinkle with salt and pepper to season while cooking; drain grease.
Stir in undrained corn, carrots, tomatoes, garlic, and onion powder; bring to a boil.
Reduce heat and simmer, uncovered, for 5 minutes, stirring occasionally.
Meanwhile, in a large saucepan, prepare 4 servings of instant mashed potatoes per package instructions.
Spoon mounds of potatoes on top of meat mixture and sprinkle potatoes with cheese.
Cover skillet and heat 2-3 minutes until cheese melts.

Use Your Tool
You may substitute frozen veggies if desired.

More Bearable Meal Suggestions
Serve with *Cheesy Cheddar Biscuits* (page 89).

2 (15-ounce) cans chili with beans
1 (16-ounce) package beef frankfurters
10 (8-inch) flour tortillas
1 (8-ounce) package extra-sharp cheddar
    cheese, shredded

Preheat oven to 425 degrees.
In a 9 x 13-inch baking dish, cover bottom with 1 can of chili.
Roll frankfurters inside tortillas and place each, side-by side, on top of chili bed seam sides down.
Pour the second can of chili over the wrapped frankfurters and sprinkle with cheese.
Cover dish with aluminum foil and bake for 30 minutes.

CHILI DOG CASSEROLE

Bring an outdoor cookout indoors! This is especially important to serve during the dead of winter when you can only dream of eating outside. I also refer to this as "Bear Scout" casserole.

109

A TRIBUTE
Say Cheese

You may have already noticed that I love cheese. I won't deny it. In fact, my European ancestry forbids me to NOT include it in my regular diet. I was taught to understand and appreciate its versatility as well as its importance in dynamic, flavorful cooking. Although this book is geared toward novice and intermediate cooks, I still encourage experimentation and the art of making food dishes uniquely your own. Cheese is a perfect food for attempting your own style of creative cooking.

If you like eating and cooking with cheeses, it is likely that you have already found a particular few that you prefer. For those of you who feel adventurous, consider the following selection before you reach for the cheddar or Swiss during your next supermarket visit. The whole cheese world is there for the tasting. Go for it!

Gorgonzola
This Italian-style blue cheese has a full, earthy flavor and is delicious blended into hot pasta or rice—instead of cheddar or Parmesan.

Gouda
This exceptionally versatile cheese is a great substitute for cheddar whether sliced on sandwiches or cubed for party trays. Try brick cheese as another tasty alternative on sandwiches.

Ricotta
Commonly introduced to Bear palettes through the classic lasagna recipe. Its creamy, granular texture offers a mild flavor with a hint of sweetness. This makes it especially suited for dessert recipes like cheesecake. See Rum Raisin Dessert (page 214).

Brie
A classic soft French cheese with an edible rind that comes in a variety of flavors. It is especially tasty when blended into soups and sauces. Ooh la la!

Feta
The definitive cheese from Greece—feta lovers already know how delicious and versatile this cheese can be. If you want to experiment with this fabulous cheese (beyond the traditional Greek salad), try it in scrambled eggs.

LEMON SHRIMP WITH PASTA

Variety is essential to a Bear diet. Occasionally one must venture out of the realm of grilled meats and explore. This is a great way to incorporate seafood into your carbo-loading regime—especially for Muscle Bears. This microwave version makes it so quick and easy!

4 cups favorite style pasta
1/2 pound fresh or frozen uncooked medium
    shrimp in shells
1 medium zucchini (1 1/2 cups),
    cut into 1/4-inch thick slices
1 yellow summer squash (1 1/2 cups),
    cut into 1/4-inch thick slices
1 green bell pepper, cut into 1/4-inch strips
1 small carrot, peeled and cut into thin slices
1 cup chicken broth
1/4 cup lemon juice
2 tablespoons cornstarch
1/4 teaspoon salt
1 teaspoon dried dill weed

In a large saucepan, cook pasta according to package instructions.

Meanwhile, peel shrimp (if shrimp are frozen, do not thaw; peel in cold water). Make a shallow cut lengthwise down back of each shrimp; wash out vein.

In a 3-quart microwave-safe casserole dish, mix shrimp, zucchini, squash, bell pepper, and carrots.

Cover tightly and microwave on HIGH for 9 minutes, stirring once after 4 minutes, until shrimp is pink and firm. Drain and let stand for 5 minutes.

In a large microwave-safe measuring cup or medium-sized bowl, blend broth, lemon juice, cornstarch, salt, and dill until smooth. Microwave uncovered on HIGH for 4 minutes, stirring after every minute, until mixture boils and thickens.

Stir in shrimp mixture; pour over pasta and serve.

Use Your Tool

A food processor will shorten prep time when cutting the vegetables.

More Bearable Meal Suggestions

Rotini is my favorite pasta for this recipe. However, linguine, capellini, or spaghetti work equally as well.

1 pound lean ground beef
1 (15-ounce) can mixed vegetables
1 (15-ounce) can cream of mushroom soup
1 (15-ounce) can cream of chicken soup
½ cup fresh mushrooms, chopped
1 (20-ounce) package frozen Tater Tots
½ cup Swiss cheese, shredded

Preheat oven to 350 degrees.
In a large skillet over medium-high heat, brown ground beef completely; drain grease and spoon ground beef into the bottom of a 2-quart casserole dish.
Drain vegetables and add to top of beef.
In a mixing bowl, blend soups and mushrooms; spread over vegetables.
Pour frozen Tater Tots on top.
Bake for 45 minutes.
Remove from oven and sprinkle cheese on top.
Let casserole cool for 5 minutes before serving.

More Bearable Meal Suggestions
Serve with *Tuxedo Bread* (page 25).

TATER TOT CASSEROLE

Sometimes comfort food appreciation can be downright inspiring.
This casserole epitomizes Bear comfort food.
Enjoy!

MY SUCCOTASH WISH

Another dee-lightful veggie recipe that suggests canned vegetables instead of frozen. If you choose to use frozen, make sure they are thawed prior to cooking.

8 ounces smoked sausage, halved lengthwise and sliced
½ cup yellow onion, chopped
1 (15-ounce) can green lima beans, drained
1 (15-ounce) can whole kernel corn, drained
½ teaspoon minced garlic
1 (14-ounce) can diced tomatoes
Salt and freshly ground black pepper to taste

Coat large saucepan with nonstick cooking spray or a half-teaspoon of oil.
Add sausage and onion and cook over medium-high heat for approximately 5 minutes until sausage is browned and onion is tender.
Add lima beans, corn, garlic, and undrained tomatoes. Reduce heat to medium low and simmer, uncovered, for 10 minutes, stirring occasionally until liquid is nearly evaporated.

More Bearable Meal Suggestions
Eliminate the sausage and serve as a side dish!

2 pounds stew meat
2 medium red onions, cut into large pieces
2 celery stalks, chopped
4 carrots, chopped into large pieces
2 cups tomato juice
1/3 cup tapioca
1/4 teaspoon freshly ground black pepper
1 tablespoon basil, chopped
2 tablespoons sugar
2 medium potatoes, cut into large pieces

Preheat oven to 350 degrees.
In a large baking pan or casserole dish, combine meat, onion, celery, and carrots.
In a separate bowl, combine tomato juice, tapioca, pepper, basil, and sugar.
Pour mixture over meat and vegetables.
Cover and bake for 2 1/2 hours.
Add potatoes, cover again, and bake for 1 additional hour.

OVEN STEW

Now this is Bear cookin', folks! So hearty, so delicious! The hardest part about this recipe is the wait. However, the slow baking makes this dish so flavorful.

115

VINEYARD PASTA

My Italian friends have taught me a thing or two about pasta appreciation. One thing I've learned is: simple is better. My favorite way of enjoying pasta with other foods has always involved simple preparation with simple, flavorful ingredients. This is a great example and one of my favorites.

1 tablespoon olive oil
1 yellow or white onion, diced
2 teaspoons minced garlic
1 pound Italian link sausage,
    sliced 1/8-inch thick
1 pound penne pasta
1 cup red wine
3 large fresh tomatoes, cored and chopped
1/2 cup basil leaves, sliced
1 tablespoon dried parsley flakes
Salt and freshly ground black pepper to taste
1/2 cup Parmesan cheese

Heat a large pot of water to a boil.
Heat olive oil in a large skillet over medium heat.
Add onion and cook approximately 3 minutes until translucent.
Stir in garlic and sausage. (If using fresh sausage, cook until meat is no longer pink, approximately 8 minutes; drain.)
Meanwhile, cook pasta in the boiling water according to package directions.
Add red wine to skillet.
Cook approximately 5 minutes until about 1/4 cup liquid remains.
Stir in tomatoes and cook for 5 minutes until softened.
Remove skillet from heat until pasta is through cooking.
Stir basil and parsley into sauce; season with salt and pepper.
Serve over drained pasta and garnish with cheese.

More Bearable Meal Suggestions
Serve with more of your favorite red wine and delicious Italian bread.

2 cups dried rotini pasta
1 ½ cups fresh mushrooms, sliced
½ cup yellow or white onion,
        coarsely chopped
1 cup cooked ham, cubed
1 (15-ounce) can cut green beans, drained
1 cup bottled ranch salad dressing
2 teaspoons Dijon mustard

Cook pasta according to package directions; drain.
Meanwhile, coat a large skillet with nonstick cooking
spray or a teaspoon of oil.
Heat skillet over medium-high heat.
Add mushrooms and onions and cook until tender,
stirring occasionally.
Reduce to medium-low heat and stir in the remaining
ingredients and heat for approximately 2 minutes.
Add the cooked pasta and heat for 1 minute.

Use Your Tool
A wok can be used instead of a large skillet, but be
careful. Wok cooking requires more attention and
more stirring while foods cook.

More Bearable Meal Suggestions
Serve with your favorite hot sliced bread and assorted
sliced cheeses.

WEEKDAY STIR-FRY

*Oh, sure. You
could come home
exhausted from
work and convince
yourself not to
cook a decent
meal, but why
deprive yourself?
Consider this one
a reward for a
hard day's work.*

TURKEY SALAD DELUXE

This one-bowl wonder is another great way to finish a Thanksgiving turkey—or enjoy it year-round.

¼ pound fresh mushrooms, thinly sliced
1 teaspoon Dijon mustard
2 tablespoons white wine vinegar
6 tablespoons olive or vegetable oil
Salt and freshly ground black pepper to taste
2 cups cooked turkey, diced
1 small red bell pepper, seeded
        and cut into strips
4 ounces Swiss cheese, chopped
½ cup cooked cut green beans
1 head iceberg lettuce
½ cup green seedless grapes, halved
¼ cup walnuts, coarsely chopped

In a large mixing bowl, combine mustard, vinegar, salt, pepper, and oil.
Add mushrooms to mixture and stir.
Set aside for 1 hour, stirring occasionally.
Add turkey, red bell pepper, cheese, and green beans to mushroom mixture and toss lightly.
Refrigerate until ready to serve.
To serve, line a large platter or deep serving bowl with lettuce leaves.
Spoon turkey mixture over lettuce and garnish with grapes and nuts.

More Bearable Meal Suggestions
This is also a great chicken salad recipe! Enjoy with a tall glass of *Sweet Iced Tea* (page 71).

1 pound (approx. 5 cups) frozen
      shoestring-cut potatoes
1 cup pizza sauce
1 teaspoon dried oregano
1/2 green bell pepper, coarsely chopped
2 tablespoons green onion, chopped
1/2 cup fresh mushrooms, coarsely chopped
2 cups mozzarella cheese, shredded
1/2 cup Parmesan cheese, grated

Preheat oven to 450 degrees.
Coat a 12-inch round pizza pan with nonstick cooking spray.
Arrange frozen potatoes evenly in one layer in pan.
Bake 10 minutes.
Remove pan from oven and cover potatoes with sauce.
Sprinkle evenly with remaining ingredients and return to oven for 15 minutes until top is lightly browned.

FRENCH FRY PIZZA PIE

Riddle: What is a hungry Bear's favorite question? Answer: Do you want fries with that pizza? Well, now you can have it all in one spectacular dish. This deserves a big "WOOF!"

119

CHICKEN SAUSAGE SUPPER

I love chicken sausage. It's a nice change of pace from the traditional beef and pork varieties. This is another recipe that I enjoy making during the week with the help of my handy microwave oven.

4 medium-sized potatoes, peeled and cut into
    1-inch cubes
1 tablespoon vegetable oil
1 yellow or white onion, coarsely chopped .
12 ounces fully cooked chicken sausages,
    cut into ½-inch slices
2 small green apples, cored and
    cut into ½-inch wedges
½ cup water
2 tablespoons Dijon mustard
1 cup milk
Salt and freshly ground black pepper to taste

Place potatoes in a microwave-safe dish; cover with plastic wrap, vent slightly.
Microwave on HIGH for 7-9 minutes or until just tender; set aside.
Meanwhile, in a large skillet, heat oil over medium-high heat.
Add onion and cook for 3 minutes, stirring occasionally.
Add sausage and cook for an additional 4 minutes or until lightly browned.
Add apples and cook an additional 3 minutes.
Add water and mustard; reduce heat to medium low.
Simmer until liquid is reduced by half; season with salt and pepper.
Mash potatoes; gradually beat in milk to reach a soft consistency.
Season with salt and pepper if desired.
Cover and reheat briefly in microwave oven, if needed.
To serve, spoon potatoes onto plate and top with sausage mixture.

Use Your Tool
To save time, I prefer to keep the lumps in my mashed potatoes!

120

1 pound lean ground beef
1 cup water
1 (6-ounce) can tomato paste
½ cup cheddar cheese, shredded
¼ cup mozzarella cheese, shredded
1 package sloppy joe seasoning mix
1 (1 pound) box elbow macaroni
1 (15-ounce) can whole kernel corn, drained
1 small yellow onion, chopped
1 small green bell pepper, seeded and chopped
Salt and freshly ground black pepper to taste

Heat a large pot of water to a boil.
In a large skillet over medium-high heat, cook ground beef thoroughly; drain grease.
Add water, tomato paste, cheese, and seasoning mix. Simmer for approximately 5-7 minutes; stirring occasionally.
Meanwhile, cook pasta in the boiling water according to package directions; drain well.
Return pasta to pot and add meat mixture, corn, onion and bell pepper; stir well and serve.

More Bearable Meal Suggestions
Serve with a fresh green salad and piping hot, buttered dinner rolls.

SLOPPY JOE CASSEROLE

Whoever Joe was (and I'm certain he was a Bear), I'm glad he was so sloppy or we might never have enjoyed this delectable, stovetop classic. Thanks, Joe.

121

BOO BOO'S BISCUIT BAKE

You won't find this in your average pic-a-nic basket. I especially like serving it on cold winter evenings.

1 (28-ounce) or 2 (16-ounce cans) baked beans
1 pound ground beef
1 small red onion, diced
½ cup green bell pepper, seeded and diced
1 tablespoon vegetable oil
1 cup milk
1 tablespoon all-purpose flour
1 tablespoon salt
Freshly ground black pepper to taste
1 (10-ounce) can refrigerated biscuit dough

Preheat oven to 450 degrees.
In a large nonstick skillet over medium-high heat, cook ground beef with onions until meat is thoroughly cooked.
Reduce heat to medium low and add baked beans, milk, flour, salt, and pepper.
Simmer for approximately 5 minutes until sauce has thickened.
Season again to taste.
Pour mixture into an 8 x 8-inch baking dish. Place uncooked biscuits on top of mixture and bake 10-12 minutes or until biscuits are lightly browned.

More Bearable Meal Suggestions
Serve with hearty ale.

At one time or another in our lives, the question of food spoilage has reached all of us. It usually occurs when you open an unrecognizable container in the back of your refrigerator and ask the question once immortalized by comedian George Carlin, "Is it meat, or is it cake?"

Despite the lack of Internet information regarding specific rates of food spoilage for particular foods (due in part to liability concerns), the U.S. government, through the Department of Agriculture and the Food and Drug Administration, offers some general rules for food safety:

- Between 40°F and 140°F (5°C and 60°C) bacteria multiply rapidly, so food should not linger in that temperature range. It should either be in the refrigerator or freezer, or being heated in an oven. Food should not be out for more than two hours (one hour if the room temperature is above 90°F).

- Do not keep food if it has been standing out for more than two hours. Do not taste it, either!

- Date leftovers so they can be used within a safe time period. Generally, they remain safe when refrigerated for three to five days. It's best to store leftovers in a plastic container that can be tightly sealed.

- Reheat foods thoroughly to 165°F (75°C), or until hot and steaming. Bring gravy and sauce to a rolling boil. Carved meat or poultry can be sprinkled with broth (to stay moist, if desired) and reheated in an oven set no lower that 325°F (160°C) or in a microwave oven.

For centuries cooks have taken the most logical approach to questioning food life. They smell and look at it. Unfortunately, you can't always see or smell that food is not "good." However, you can usually tell with this method when it has outlived its usefulness. Beyond this, simply use common sense. When in doubt, throw it out!

DID YA KNOW?

Leftover Life

123

CHICKEN DIVINE

This is the kind of chicken dish that frozen-food manufacturers attempt to duplicate. Unfortunately, they can't compete with the freshness and satisfaction of this recipe.

3 cups cooked chicken breast,
    cut into large pieces
1 pound fresh broccoli
1 tablespoon melted butter
1 cup chicken broth
1 (10-ounce) can cream of mushroom soup
1 cup sharp cheddar cheese, grated

Preheat oven to 350 degrees.
Cut broccoli heads from bunch keeping partial stems attached.
In a large saucepan, bring water to a boil; add broccoli and cook for approximately 6-8 minutes and drain thoroughly.
In a 9 x 13-inch baking dish, place chicken in a single layer.
Top with broccoli and butter.
In a separate bowl, blend broth, soup, and cheese.
Pour this mixture evenly over broccoli and chicken.
Bake for 20-25 minutes.

Use Your Tool
Another worthy reason to invest in a versatile 9 x 13-inch baking dish!

124

1 ½ cups all-purpose flour
2 cups whole milk
2 eggs
8 ounces Muenster cheese, chopped
4 ounces pepperoni, thinly sliced and chopped
Dash of each: salt, freshly ground black pepper,
    oregano

Preheat oven to 350 degrees.
In a large mixing bowl, combine flour, milk, eggs, cheese, and pepperoni.
Pour mixture into a greased 9 x 13-inch glass baking dish.
Sprinkle with salt, pepper, and oregano.
Bake for 30 minutes or until top is golden brown.
Remove from oven and let cool for at least 5 minutes.

More Bearable Meal Suggestions
Serve with a hearty side salad and your favorite beer.

PEPPERONI PIE

I am so addicted to this recipe. It has turned me off frozen pizzas forever! There is something magical about the way all of the ingredients bake together. One taste and you'll be hooked.

125

COME-AND-GET-IT
ENTRÉES

BEAR CHOPS & BEANS

This stick-to-your-ribs classic is a year-round favorite. Serve this and I promise the Bears will come running!

2 (16-ounce) cans favorite flavor/brand
　　of baked beans
4 lean pork chops, cut 3/4-inch thick
1 tablespoon vegetable oil
1/4 cup yellow or white onion, finely chopped
1 (8-ounce) can whole kernel corn, drained
1/2 cup celery, thinly sliced
1 envelope dry onion soup mix
2 tablespoons steak sauce
Freshly ground black pepper to taste

In a large skillet over medium-high heat, brown pork chops in hot oil on each side for several minutes; drain grease.
In a separate bowl, combine remaining ingredients and pour over chops.
Bring to a boil. Cover and reduce heat; simmer for 30 minutes or until chops are tender.

Use Your Tool
If your skillet isn't large enough, cut chops from the bone after browning to make more room.

More Bearable Meal Suggestions
Baked bean brands have improved their flavor selection considerably over the years. This recipe gives you the chance to experiment beyond the standard flavor. Go for it!

126

4 tablespoons olive oil
8 ounces Italian sausage, casings removed
1 cup eggplant, diced ½-inch thick
1 cup zucchini, diced ½-inch thick
1 cup red bell pepper, diced ½-inch thick
⅓ cup yellow onion, chopped
2 teaspoons minced garlic
1 (28-ounce) can Italian plum tomatoes,
    drained and chopped
2 tablespoons fresh parsley, chopped
2 teaspoons fresh basil leaves, chopped
Freshly ground black pepper to taste
3 ounces mozzarella cheese, grated

Preheat oven to 350 degrees.
In a large saucepan over medium heat, cook sausage thoroughly for approximately 10 minutes, breaking meat into pieces using a wooden spoon or utensil.
Place meat into separate bowl and set aside.
Drain most of the oil from pan.
In the same saucepan over medium heat, add eggplant, zucchini, bell pepper, onion, and garlic.
Cook 10-15 minutes or until softened, stirring occasionally.
Stir in reserved sausage, tomatoes, parsley, basil, and pepper.
Reduce heat and simmer for 15 minutes.
Spoon mixture into an 8 x 8-inch baking dish and sprinkle with cheese.
Bake for 15-20 minutes or until cheese melts.

ITALIAN SAUSAGE CASSEROLE

So many European countries are known for their sausages. However, my favorite varieties always come from Italy. This casserole allows you to truly appreciate its wonderful flavor. And just wait until the aroma fills your kitchen. "Mamma Mia!"

127

SAN FRAN CHICKEN CASSEROLE

This takes the world-famous "San Francisco treat" and glorifies it. For me, this is classic casserole cooking.

4 boneless, skinless chicken breasts
1 (10-ounce) can cream of chicken soup
1 (10-ounce) can cream of mushroom soup
1 (10-ounce) can cream of celery soup
1 cup whole milk
1/4 cup Catalina-style salad dressing
2 packages Rice-A-Roni,
        excluding seasoning packets

Preheat oven to 350 degrees. In a large mixing bowl, combine rice mixture (do not use seasoning packets), soups, milk, and salad dressing; blend well.
Pour mixture into 9 x 13-inch baking dish.
Place chicken breasts on top and spoon rice mixture onto chicken to cover.
Cover dish with aluminum foil and bake for 60-70 minutes.

More Bearable Meal Suggestions
Serve with *Easy Southern-style Greens* (page 92).

128

1 (10-ounce) can cream of potato soup
³/₄ cup whole milk
¹/₈ teaspoon freshly ground black pepper
1 cup frozen mixed vegetables, thawed
1 cup cooked chicken breast, chopped
1 egg
1 cup all-purpose baking mix (Bisquick)

Preheat oven to 400 degrees.
In a mixing bowl, blend soup, ¼ cup of milk,
vegetables, and chicken.
Pour into a 9-inch pie pan or baking dish.
In the same bowl, blend remaining milk, egg, and
baking mix.
Pour this mixture over the chicken mixture.
Bake for 30 minutes or until golden brown.

More Bearable Meal Suggestions
Serve with *Tangy Coleslaw* (page 96).

CHEATER'S CHICKEN POT PIE

I like traditional homemade pot pies as much as the next Bear, but c'mon. I want it NOW! So here's a much faster version that's just as satisfying.

A TRIBUTE
Mayonnaise

There are certain things in this world that I have always known to be truth: the love of my family, the wonderful feeling of beach sand between my toes, the joy of Christmas morning, and my unending love of mayonnaise.

Perhaps my appreciation stems from my Southern upbringing, perhaps it's from my French blood. Whatever the reasons, I am grateful that I am not one of "those people" who do not have an affinity toward this remarkable food invention.

Theories abound regarding its origin—from Roman and Egyptian times to sixteenth-century Europe. The most common belief is that mayonnaise was invented in 1756 by the French chef of the Duc de Richelieu. After the Duc beat the British at Port Mahon, his chef created a victory feast that was to include a sauce made of cream and eggs. Realizing that he had no cream, the chef substituted olive oil for the cream and "voila!" The chef named the new sauce "Mahonnaise" in honor of the Duc's victory.

What makes it so remarkable is that it's an emulsion. The emulsifier is egg yolk, which contains the fat emulsifier lecithin that keeps the lemon juice or vinegar and the egg yolks from separating. Oil is carefully added to thicken and seasoning is added for more flavor. The science of its ingredients makes it one of the most versatile food products for cooking and baking. It can be found as a base in sauces, spreads, and dips, also in salads, casseroles, cakes, and breads to name a few. Frankly, I prefer to simply slather it on my favorite sandwiches and enjoy.

Now you can find a new variety of mayonnaise flavors in many gourmet markets including cilantro, basil, citrus, or aioli, the traditional garlic mayonnaise. If you're really adventurous in the kitchen, try making your own, but beware. Homemade mayo lasts only a few days, commercially made brands can last months.

POTATO SAUSAGE SUMMER SALAD

*For all of you
"meat n' potatoes"
lovers, this is
definitely for you.
It's especially
impressive for
dinner parties.*

2 pounds small red potatoes, cooked,
    cooled, sliced
3 tablespoons white wine vinegar
1/3 cup olive oil
1/2 teaspoon salt
1/4 teaspoon freshly ground black pepper
1 1/2 pounds smoked sausage, grilled or
    cooked and cut into 1/2-inch pieces
1/3 cup green onions, sliced
1 tablespoon dried parsley flakes
Lettuce leaves

In a large mixing or serving bowl, gently toss cooked
potatoes (preferably cooled to room temperature),
vinegar, oil, salt, and pepper.
Add sausage to potatoes and blend.
Sprinkle with green onion and parsley.
To serve, place large lettuce leaves on each plate and top
with salad mixture.

More Bearable Meal Suggestions
Serve with your favorite beer and bottle of merlot.

2 tablespoons green onions, finely chopped
2 tablespoons lime juice
1/4 teaspoon salt
2 cups cooked chicken breast, chopped
1 cup cooked green peas
1 cup mayonnaise
1/4 cup carrots, finely chopped
1/4 cup celery, finely chopped
3 tablespoons orange juice
1/2 teaspoon salt
1/2 teaspoon ground cinnamon
1/4 teaspoon freshly ground black pepper

Garnish
Lettuce leaves
2 navel oranges, pared and cut into
    thin wedges

In a small bowl, sprinkle green onions with lime juice
and 1/4 teaspoon salt; cover and refrigerate.
In a larger bowl, combine the remaining ingredients
excluding the garnish.
To serve, place large lettuce leaves on each plate and top
with salad mixture.
Add orange wedges to sides of salad servings and
sprinkle with green onions.

Use Your Tool
A food processor will help speed the vegetable chopping
time.

More Bearable Meal Suggestions
Serve with *My Favorite Deviled Eggs* (page 95) and *Sweet
Iced Tea* (page 71).

ORANGE CHICKEN SALAD

*During the
summer months,
I sometimes double
the recipe so I can
enjoy it twice as
long.*

133

(I'M) NUTS OVER CHICKEN AND BISCUITS

I first sampled this little taste of heaven many years ago when trapped by snow in a mountain cabin with friends. We held a "Last Meal" cooking contest with our remaining provisions. This was my entry and it won! (Our rescuers found us very full.)

2-3 pound broiler-fryer chicken,
   cut into whole pieces
1 envelope dry onion soup mix
1 (10-ounce) can cream of mushroom soup
2/3 cup water
1 (4-ounce) can mushrooms, drained
2 cups all-purpose baking mix (Bisquick)
2/3 cup whole milk
1 teaspoon dried parsley flakes
1/2 cup pecans, chopped
Paprika

Preheat oven to 400 degrees.
Place chicken, skin side up, in an ungreased 9 x 13-inch baking dish.
Reserve 2 tablespoons of onion soup mix (dry).
In a mixing bowl, add remaining dry soup mix, cream of mushroom soup and water; pour over chicken.
Cover and bake for 60 minutes.
Remove from oven and spoon mushrooms over chicken.
In a mixing bowl, blend baking mix, milk, reserved dry onion soup mix, and parsley flakes until soft dough forms; beat vigorously for 30 seconds.
Drop spoonfuls of dough onto chicken; sprinkle with paprika and nuts.
Bake uncovered approximately 12 minutes or until biscuits are lightly browned.

Use Your Tool
An electric mixer is ideal when creating dough.
Otherwise, the old-fashioned way works as well.

134

4 large baking potatoes, fully cooked
1 tablespoon butter
1 medium yellow onion, sliced widthwise
     into 1/2–inch rings
Dash of salt and freshly ground black pepper
1 1/2 cup extra-sharp cheddar cheese, shredded
1/2 cup cooked ham, chopped
1/2 cup cooked broccoli heads, chopped

In a large skillet over medium-high heat, sauté onions in melted butter, salt, and pepper until onions are tender and appear translucent; add to a large mixing bowl.
Split each potato lengthwise and scoop out center portion of potato; add this to bowl with cooked onions.
To same bowl, add cheese, ham, and broccoli.
Mix well.
Spoon this mixture back into potato shells; add additional cheese on top.
Microwave on HIGH for 2-3 minutes.

Use Your Tool
Get creative—substitute turkey, chicken, or ground beef!

BEAR PIG STUFFED POTATOES

A Bear I once noticed at a buffet restaurant inspired this. He was wearing a "Bear Pig" T-shirt and eating a stuffed baked potato the size of a basketball player's shoe. I was so proud that it brought a tear to my eye.

135

CHINESE PORK AND VEGETABLES

In an "international mood"? Try stir-fry for a change of pace.

¼ cup soy sauce
1 tablespoon cornstarch
2 tablespoons vegetable oil
1 clove garlic, minced
1 teaspoon gingerroot, grated
1 pound lean pork, diced
3 cups red or green cabbage, chopped
1 (10-ounce) package frozen peas, thawed
1 green bell pepper, cored, seeded, and
       cut into thin strips
1 (3-ounce) can sliced mushrooms, drained
2 tablespoons green onions, sliced
2 tablespoons dry sherry
Hot cooked rice

In a small bowl, combine soy sauce, cornstarch, and ½ cup water; set aside.
Heat oil over medium heat in a wok or an electric skillet. Add garlic, gingerroot, and a dash of salt and cook until garlic is just golden.
Increase to high heat and add pork gradually.
Stir constantly for 2-4 minutes; season with additional salt if desired.
Add cabbage, peas, green pepper, mushrooms, green onions, and sherry.
Cook and stir for 1 additional minute.
Pour soy mixture over meat-vegetable mixture; cook and stir until thickened. Serve over rice.

Use Your Tool
Make sure that pork is diced so it will cook thoroughly. Also, don't forget to cook rice before you stir-fry! A rice steamer is ideal for keeping rice hot and moist before serving.

More Bearable Meal Suggestions
Chinese beer.

2 tablespoons butter, melted
1/8 teaspoon ground nutmeg
1/8 teaspoon ground allspice
1 (17-ounce) can sweet potatoes,
     drained and mashed
1/2 cup yellow onion, chopped
1 tablespoon butter
2 cups cooked turkey, diced
1 (10-ounce) can cream of mushroom soup
1 (8-ounce) can whole kernel corn, drained
1 (8-ounce) can peas, drained
1 small tomato, peeled and diced

Preheat oven to 350 degrees.
In a mixing bowl, beat 2 tablespoons butter, nutmeg,
allspice, and 1/2 teaspoon salt into mashed sweet
potatoes.
Line a 9-inch pie pan or dish with potato mixture to
form a shell; build edges 1/2-inch high.
In a large skillet over medium heat, cook onion in
1 tablespoon of butter until slightly tender.
Stir in turkey, soup, corn, peas, tomato, and a dash of
salt.
Spoon turkey mixture into sweet potato shell.
Bake for 35 minutes.

SWEET POTATO TURKEY PIE

*For me, everything
seems to go back
to Thanksgiving
dinner. Here is
another example
of my lifelong
obsession.*

137

BIG FAT GREEK SPINACH PIE

If you're in an "international mood", this is a modified version of the Greek dish known as spanakopita. Popeye never had it so good.

2 bunches green onions, chopped
¼ cup olive oil
3 pounds fresh spinach, washed,
        trimmed, and chopped
1 pound feta cheese, crumbled
4 eggs, beaten
½ cup minced parsley
Dash of salt and freshly ground black pepper
12 sheets (½ package) phyllo dough
1 cup butter, melted

Preheat oven to 350 degrees.
In a skillet over medium heat, brown green onions in olive oil until tender.
In a large bowl, mix spinach, cheese, eggs, and parsley; add cooked green onions, salt and pepper. Mix well.
Line a greased 9 x 13-inch baking pan with 6 phyllo sheets, brushing each sheet with melted butter.
Spead spinach mixture over phyllo.
Cover with remaining phyllo sheets, brushing each sheet with butter.
Bake 45 minutes.
Cool and cut into squares.

More Bearable Meal Suggestions
This may be served hot or cold. Cut into smaller squares if you want to serve with white wine as an appetizer; larger squares if served as a side dish or entrée.

138

1 (14-ounce) package macaroni and
　　cheese dinner (KRAFT Deluxe brand
　　most recommended)
1 pound lean ground beef
1 (14-ounce) can diced tomatoes in juice
2 tablespoons chili powder

Cook macaroni and cheese dinner according to package directions.
Meanwhile, in a large skillet, cook ground beef thoroughly; drain.
Add beef and remaining ingredients to prepared macaroni and cheese; heat and blend stirring occasionally.

More Bearable Meal Suggestions
For an added touch, sprinkle crumbled corn chips on top of each serving! Serve with beer or sangria.

SUPER DELUXE CHILI MAC

A book about comfort foods wouldn't be complete without this. Here is a super-easy version of this hearty, Bear-worthy classic.

139

Section V

MORE BEAR MEAT

BARBEQUE MEAT LOAF

Meat loaf, as a comfort food, is as American as baseball, apple pie, and...Bear lovin'. It just makes sense that meat loaf should be included in this collection. I've enjoyed many over the years, but I especially like this version.

1 tablespoon vegetable oil
1 yellow onion, chopped
2 pounds lean ground sirloin beef
½ cup Italian-style bread crumbs
½ cup hickory-smoke-flavored barbeque sauce
2 large eggs
3 tablespoons ketchup
1 tablespoon steak sauce
2 teaspoons salt

Preheat oven to 325 degrees.
Heat oil in a skillet over medium heat; add onion.
Cook for approximately 10 minutes or until onion becomes tender and caramelizes.
In a large mixing bowl, add cooked onion, meat, and remaining ingredients.
Mix thoroughly (easiest with clean hands), and pack into an 8 x 4-inch loaf pan.
Place loaf pan onto a baking sheet to catch overflow in oven.
Cook 40-45 minutes.

Use Your Tool

To test meat's doneness more accurately, place an instant-read meat thermometer into center of loaf while baking. The thermometer should reach 155 degrees.

More Bearable Meal Suggestions

Serve with *Yam-pple Casserole* (page 73), *Southern-Baked Corn* (page 77), or any of the other great recipes in Section III (page 65).

MAPLE-GLAZED HAM

This glaze has a "sweet kick" to it and is perfect for a nice juicy baked ham. It also works well on other grilled meats.

1 cup maple syrup
1/3 cup honey
2 tablespoons Dijon mustard
4 tablespoons apple cider vinegar
1/4 teaspoon cayenne pepper
1 (10-12 pound) ham, bone-in

Preheat oven to 325 degrees.
In a saucepan over medium heat, combine all glaze ingredients for ham.
Bring to a boil; reduce heat and stir occasionally until liquid reduces by half.
Trim fat from ham and place in roasting pan.
Brush ham thoroughly with glaze.
Bake 90-120 minutes or until ham is cooked through.
Baste with glaze every 15-20 minutes.

More Bearable Meal Suggestions

Serve with *Sage Dressing* (page 84) and *Pumpkin Cheese Bake* (page 90), or *Summertime Carrot Salad* (page 91) and *My Favorite Deviled Eggs* (page 95).

143

EASY-BAKE SWISS STEAK

This is a favorite of my Bear-friend, Max, who devours it faster than I can say, "Pass the peas and carrots."

1 ½ pounds round steak, 1-inch thick
3 tablespoons all-purpose flour
½ teaspoon salt
1 teaspoon dry mustard
¼ teaspoon paprika
1 (16-ounce) can stewed tomatoes
¼ teaspoon salt
¼ teaspoon freshly ground black pepper
2 tablespoons Worcestershire sauce

Preheat oven to 350 degrees.
Trim excess fat from steak; reserve fat.
In a small bowl, mix flour, salt, mustard, and paprika.
Rub half of the mixture on the steak; then pound steak thoroughly with a meat tenderizing mallet.
Turn steak and repeat steps with other side.
In a large skillet over medium-high heat, melt a few ounces of trimmed fat; add steak and brown both sides thoroughly.
In a mixing bowl, blend tomatoes, salt, pepper, and Worcestershire sauce.
Place browned steak in a large baking dish; pour tomato mixture over steak. Bake for 75 minutes or until meat is fork tender.

Use Your Tool
Some cooks use a saucer instead of a mallet. However, a mallet is a worthy kitchen-accessory investment.

More Bearable Meal Suggestions
Serve with *Bacon & Buttermilk Mashed Potatoes* (page 66).

144

½ cup all-purpose flour
½ cup yellow cornmeal
2 teaspoons salt
1 teaspoon ground red pepper
½ teaspoon freshly ground black pepper
½ teaspoon dried parsley flakes, crushed
2 ½ pounds fish fillets
½ cup vegetable or peanut oil
4 tablespoons green onion, minced

In a small bowl, blend flour, cornmeal, salt, red and black pepper.
Spread this mixture on a large plate.
Dip fillets into flour mixture, shaking off excess; transfer floured fillets to a clean plate.
Heat oil in a large skillet over medium heat.
Fry fish, turning once, until golden brown on both sides, approximately 8 minutes.
Transfer fried fish to paper towels to drain.
Garnish with green onion before serving.

Use Your Tool
As always, please use caution when frying with oil.

More Bearable Meal Suggestions
Serve with lemon wedges and your favorite brand of tarter sauce. Also serve with *Easy Southern-style Greens* (page 92) and *Potato Salad from Hell* (page 99).

EVEN I CAN FRY FISH

It's as easy as the name suggests! I recommend frying cod, flounder, or orange roughy for the best flavor and texture results.

FRIED TURKEY

Always eating ordinary fried chicken? Think outside the box, man.

1 cup dry bread crumbs
¼ cup Romano cheese, grated
2 teaspoons Italian seasoning
1 cup heavy cream
1 pound boneless, skinless turkey breast halves, cut in half
¼ cup vegetable or peanut oil

In a shallow bowl, combine bread crumbs, cheese, and seasoning.
Pour cream into a separate bowl.
Dip turkey in cream then in crumb mixture.
Heat oil in a large skillet over medium heat.
Cook turkey in oil, turning once, approximately 8–10 minutes or until golden brown and juices run clear.

More Bearable Meal Suggestions
Serve with *Shocking Pink Cranberry Relish* (page 82) and *Pumpkin Cheese Bake* (page 90).

146

SLOW-COOKED BEEF & SAUSAGE LOAF

Break out the slow cooker and get ready for some serious meat eating!

1 pound lean ground beef
1 (4-ounce) can mushrooms with liquid
1/2 pound ground pork sausage
1 large egg, beaten
2 cups white sandwich bread, chopped
    into 1/2-inch pieces
1/2 teaspoon onion salt
1 cup celery (approximately 2 stalks),
    finely chopped
1/2 teaspoon garlic salt
1 tablespoon horseradish
1 teaspoon dried parsley flakes

In a large mixing bowl, combine all ingredients thoroughly (easiest with clean hands); form into rounded loaf.

Place into slow cooker; cook on LOW setting for 9-10 hours.

To test doneness, puncture loaf; juice should run clear.

Use Your Tool

It is most convenient to mix loaf in the evening and refrigerate overnight. Begin cooking loaf the following morning so it will be ready for dinner.

More Bearable Meal Suggestions

Serve with any kind of potato—preferably mashed with gravy or scalloped.

147

EASIEST CHICKEN BAKE

This is so much better than the premade "shake-in-the-bag-and-bake" stuff—honest! Besides, how can you go wrong when mayonnaise is involved?

1 pound boneless, skinless chicken breast halves
½ cup mayonnaise
1 cup Italian-style bread crumbs
¼ cup Parmesan cheese, grated

Preheat oven to 425 degrees.
Brush chicken on all sides with mayonnaise.
Blend bread crumbs and cheese in a large bowl.
Dip chicken pieces into crumb mixture and arrange chicken onto baking sheet.
Bake 20 minutes or until chicken is golden brown and thoroughly cooked.

Use Your Tool
Don't forget to line your baking sheet with foil for easier clean up!

HONEY, LICK MY CHOPS

The title of this summertime favorite speaks for itself.

2 tablespoons honey
1 tablespoon lime juice
1 tablespoon coarsely ground black pepper
½ teaspoon dried minced onion
Salt to taste
4 thin boneless pork chops

In a shallow dish, combine juice, honey, pepper, onion, and salt.
Add pork chops; turn to coat; cover and refrigerate for at least 30 minutes.
Heat broiler.
Lightly oil broiler pan.
Remove pork from marinade and pat dry.
Discard marinade.
Broil pork, turning once, until cooked thoroughly, approximately 8–10 minutes.

Use Your Tool
These chops can also be grilled outdoors.

More Bearable Meal Suggestions
Want a reason to make these? I recommend hosting a Hawaiian-themed dinner party in the dead of winter.

149

PEANUT BUTTER BEEF

I always encourage friends to have fun and be inventive when cooking with food products. This is an easy and clever way to incorporate peanut butter into a meal. I love the combination of flavors!

1 tablespoon minced garlic
1 pound stew beef, cut into thin strips
1 cup white onion, sliced
1 cup celery, chopped
2 tablespoons vegetable or olive oil
2 tablespoons peanut butter, creamy or
   chunky style
2 tablespoons soy sauce
1/2 teaspoon sugar
1 cup beef stock or bouillon
Freshly ground black pepper to taste

In a large skillet over medium heat, sauté garlic, beef, onion, and celery in oil until lightly browned.
Add remaining ingredients and bring to a boil.
Cover and reduce heat; simmer for 50-60 minutes.
Add additional beef broth or water during cooking; stir occasionally.
Serve over rice or noodles.

More Bearable Meal Suggestions
Serve with *Cheesy Cheddar Biscuits* (page 89).

DEEP-FRIED CHICKEN CRUNCH

Cereal for dinner? Why not? Besides, once you sample this yummy homemade fried chicken, you'll want it for breakfast as well!

1 cup Cap'n Crunch cereal
1 cup cornflakes cereal
1 cup all-purpose flour
1/4 teaspoon dried parsley flakes, crushed
1 tablespoon minced onion
1 tablespoon minced garlic
2 teaspoons freshly ground black pepper
1 cup whole milk
2 eggs, beaten
1 pound boneless skinless chicken breast meat,
	cut into thin strips
Vegetable or peanut oil

In a blender or food processor, pulse cereals until consistency of bread crumbs; pour into bowl.
In a second bowl, blend flour, parsley, onion, garlic, and pepper.
In a third bowl, blend egg and milk.
Dredge chicken pieces first in egg wash, then in flour mixture, and lastly in cereal mixture.
Cook chicken in 325-degree oil using deep fryer or pot for 3-4 minutes or until thoroughly cooked.

Use Your Tool
To help chicken stay moist when frying, coat it twice before frying.

More Bearable Meal Suggestions
Serve with an array of your favorite sauces and condiments.

151

CARIBBEAN PORK

I once asked a close friend and notable chef how Jamaicans originated the term "jerk" to describe their world-famous meat rub. In his infinite wisdom, he replied, "With their first bite, they jerked back from the table and hollered, 'Damn, that's good!'" Here is a variation of the classic that is sure to jerk you back.

1/3 cup white or yellow onion, chopped
1/2 teaspoon dried parsley flakes, crushed
1 teaspoon dried thyme, crumbled
1 teaspoon each sugar and salt
1 teaspoon each black pepper and
    cayenne pepper
1/2 teaspoon ground allspice
1/8 teaspoon each grated nutmeg and
    cinnamon
2 (4-ounces each) pork chops,
    cut 1/2-inch thick

In a mixing bowl, mash onions with remaining ingredients into a coarse paste.
Pat pork chops dry; rub jerk paste onto both sides of each chop.
Grill meat at least 6 inches above glowing coals for approximately 4 minutes on each side or until cooked thoroughly.
Chops can also be broiled for same length of time using moderately high heat.

More Bearable Meal Suggestions
Serve with *Barbequed Onion Bowls* (page 83) and *Easy Southern-style Greens* (page 92).

152

Most cooks know the importance of fats and oils in cooking and baking their favorite foods. Their use sparks many of the most common kitchen questions such as: Why do some recipes call for margarine instead of butter? What are the advantages to shortening? What's the difference between regular and "extra virgin" olive oil? Keep in mind that fats and oils behave differently for many reasons, starting with their sources and structures.

Some fats and oils come from animals, some from plants; some are natural, some are manufactured. Fats are solid at room temperature while oils are liquid. Certain rules apply when deciding on substitutions. If foods are heavily spiced or seasoned, you can consider alternative fats in recipes. However, remember to substitute solid for solid, liquid for liquid. The following guide is designed to address and define some players in the "fat game":

The Skinny on Fat

Butter The French rarely cook with anything else and for good reason. This natural fat is about 20 percent water, but the water is emulsified—suspended throughout the butter. It allows for the best flavor when baking, especially when taste is more important as in certain cakes and cookies. Otherwise, mechanically, margarine works like butter.

Margarine Another fat in which hydrogenated vegetable oil is processed to resemble butter with coloring, flavoring, and moisture added. Light butter or margarine means more water and less fat in the product—not recommended for most baking.

Vegetable shortening This fat, like margarine, is made from hydrogenated vegetable oils. The flavor is neutral.

Lard Rendered from pork fat; distinctive taste. Pioneers were known to use this beyond the kitchen—for medicinal purposes (mild skin burns) and occasionally to grease farm equipment.

Olive oil Used for centuries; known for its health benefits. Extra virgin olive oil is simply more refined and contains very little acid (about one percent). Olive oil marked "fino" (meaning fine) is simply a mixture of extra virgin and virgin olive oils.

Vegetable oil Most popular in American cooking. May come from almonds, corn, pumpkin seed, hazelnuts, cottonseed, grapeseed, peanuts, sesame seed, soybeans, or sunflowers. Great for frying.

SOUR CREAM AND ONION BURGERS

Why settle for a plain, boring ground beef patty when you can enjoy this on your grill? The answer is easy!

2 pounds lean ground beef
1 envelope dry onion soup mix
1 cup sour cream
½ cup Italian-style dry bread crumbs
⅛ teaspoon freshly ground black pepper
¼ teaspoon dried parsley flakes

In a large mixing bowl, combine all ingredients.
Mix thoroughly (easiest with clean hands); form meat mixture into palm of hands and pat back and forth into each hand until meat is packed. (This prevents patties from crumbling on grill.)
Once packed, form palm-sized patties approximately ½-inch thick.
Grill away from direct flame in medium-heat area for 10-15 minutes; turning once until center is no longer pink and juice appears clear.

Use Your Tool
Keep in mind, the leaner the ground beef the less likelihood for burger shrinkage when grilling—and we all know how much you hate shrinkage, so don't even go there.

More Bearable Meal Suggestions
Sprinkle burgers with French-fried onions when serving on buns for extra fun!

154

5 bacon slices
1 ½ pounds lean ground beef
1 large yellow onion, chopped
1 cup seasoned bread crumbs
2 tablespoons mayonnaise
1 egg
2 tablespoons sweet pickle relish
2 teaspoons dry mustard
1 teaspoon salt
½ teaspoon freshly ground black pepper
1 cup extra-sharp cheddar cheese, shredded
¼ cup ketchup

Preheat oven to 350 degrees.
In a large skillet over medium heat, cook bacon until limp (approximately 3 minutes); drain on paper towels and set aside.
In a large mixing bowl, combine ground beef, onion, bread crumbs, mayonnaise, egg, relish, mustard, salt, pepper, and cheese thoroughly (easiest with clean hands); form and pat into rounded loaf.
Place into a 9 x 13-inch baking dish.
Spread ketchup on top of loaf followed by bacon slices.
Bake 40-45 minutes until loaf is firm and bacon is crisp.
Let stand in dish 12-15 minutes before serving.

More Bearable Meal Suggestions
Serve on buns with fries on the side or with mashed potatoes, gravy, and *Nuts About Veggies* (page 81).

BACON DOUBLE CHEESE- BURGER LOAF

It's probably obvious by now that I love meat loaf. This one epitomizes Bear comfort food. If a Bear doesn't like this, tie him up, shave his fur, and tear his membership card into pieces.

155

MARINATED FLANK STEAK

Preserves, jams, and marmalades are pure heaven to me. Most people only enjoy them on breads without recognizing their versatility. Try this marinade and you'll discover it for yourself.

1 (12-ounce) jar orange marmalade
½ cup red wine vinegar
2 tablespoons Worcestershire sauce
½ cup ketchup
1 tablespoon chili powder
1 tablespoon steak sauce
2 tablespoons Dijon mustard
3 pounds flank steak
Salt and freshly ground black pepper to taste

In a large bowl, whisk together all ingredients except steak and salt and pepper.
Reserve 1 cup of marinade.
Add steak to remaining marinade.
Cover and refrigerate overnight.
In a saucepan over medium-high heat, boil the reserved marinade for approximately 5 minutes or until thickened into a glaze.
When ready to grill, remove meat from marinade and season both sides with salt and pepper.
Discard marinade.
Grill approximately 4-5 minutes on each side for a medium to medium-rare doneness.
When ready to serve, brush meat with glaze; cut against the grain into thin slices.

More Bearable Meal Suggestions
Serve with *Barbequed Onion Bowls* (page 83) and *Potato Salad from Hell* (page 99).

156

DEVILED CHICKEN LIVERS

Granted, I admit it's not for everyone. But for those who dare, watch out!

2 tablespoons all-purpose flour
2 teaspoons paprika
½ pound chicken livers
¼ cup yellow or white onion, chopped
2 tablespoons butter
¼ teaspoon each salt, black pepper, and
    cayenne pepper
½ teaspoon dry mustard
1 teaspoon Worcestershire sauce
½ cup chili sauce
½ cup water

In a small bowl, blend flour and paprika; coat livers in mixture.
In a large skillet over medium heat, sauté coated livers and onions in butter for 3-4 minutes until browned.
Stir in remaining ingredients and simmer for 1-2 minutes.
Serve over toasted bread or hot cooked rice.

157

MEAT AND POTATOES LOAF

Can you tell by now that I live in the Midwest? This should be a dead giveaway. Besides, food trends may come and go but the comfort from eating meat and potatoes will never fall from fashion.

1 ½ cups frozen shredded hash brown
        potatoes, thawed
1 cup red onion, chopped
3 eggs, beaten
½ cup Italian-style bread crumbs
⅓ cup ketchup
1 ½ teaspoons salt
½ teaspoon pepper
1 ½ pounds lean ground beef

Preheat oven to 375 degrees.
In a large mixing bowl, combine potatoes, onion, eggs, bread crumbs, ketchup, salt, and pepper; mix well.
Add ground beef and mix thoroughly (easiest with clean hands).
Form meat mixture into an 8 x 4-inch rounded loaf ; place into a broiler pan.
Bake 75 minutes or until meat thermometer reads 160 degrees and center is no longer pink.
Brush additional ketchup on top before serving, if desired.
Let stand for 5 minutes prior to slicing.

More Bearable Meal Suggestions
Serve with your favorite vegetables and hot buttered dinner rolls.

2 pounds beef boneless top round steak
    (approx. 1-inch thick)
¼ cup lime juice
3 tablespoons tequila
2 tablespoons vegetable oil
2 cloves garlic, finely chopped
½ teaspoon each salt and freshly ground
    black pepper
½ teaspoon ground cumin
½ teaspoon ground cayenne pepper

Pierce beef several times on both sides with a fork.
In a shallow dish, combine all other ingredients.
Add beef to dish and cover; refrigerate for at least 6
hours or overnight.
Heat coals or gas grill for direct heat.
Remove beef from marinade and apply to oiled grill;
cover.
Cook beef 4-5 inches from medium heat, turning
occasionally, for approximately 20 minutes for
medium- rare doneness.

Use Your Tool
If a shallow covered dish is unavailable, try marinating
in a large, sealable plastic food bag.

More Bearable Meal Suggestions
Serve with skillet-fried potatoes and Tangy Coleslaw
(page 96). And don't forget the beer!

BLITZED BEEF

You know you've matured when you've stopped entertaining your guests with tequila shooters and start serving it for dinner. This is one of the best ways and, like the Bourbon Beef Tenderloin (page 180), one of my favorite marinades.

CHICKEN GUIESEPPI

An Italian who will always hold my most sincere admiration and respect inspired this recipe. Thank you, Joe. I miss you.

4 boneless, skinless chicken breast halves
1 (14-ounce) can diced tomatoes
1 (8-ounce) can tomato sauce
½ teaspoon dried oregano
½ teaspoon garlic powder
½ cup red table wine
6 ounces fresh mushrooms, sliced

Coat a large skillet with nonstick cooking oil spray or 1 tablespoon of olive or vegetable oil.
Place chicken into skillet and brown over medium-high heat for approximately 5 minutes, turning once.
Reduce heat to medium-low setting.
Combine remaining ingredients in a separate bowl and pour over chicken.
Rearrange chicken in the sauce and simmer 10-12 minutes or until chicken is fully cooked and sauce has thickened.

More Bearable Meal Suggestions
Serve over linguine, spaghetti, or rice and a side of bread sticks.

OVEN-BAKED PORK CHOPS

1 egg
2 tablespoons water
¼ cup each seasoned bread crumbs,
    all-purpose baking mix (Bisquick),
    all-purpose flour
1 teaspoon each ground oregano, paprika,
    Italian seasoning, salt
½ teaspoon freshly ground black pepper
4 bone-in pork chops, cut 1-inch thick
3 tablespoons vegetable oil

Preheat oven to 425 degrees.
In a shallow bowl, beat egg and water; set aside.
Combine bread crumbs, baking mix, flour, and all
seasonings in a large plastic food bag.
Add chops to bag; shake to coat.
Remove chops and dip into egg mixture.
Return chops to bag and shake again to coat.
Place chops in a 9 x 13-inch baking pan coated with oil.
Place chops in pan and cook for 15 minutes; turn once.
Cook an additional 15 minutes or until cooked to
desired doneness.

Use Your Tool
To prevent overcooking your chops, make sure they are
cut at least 1-inch thick.

As I implied with the Easiest Chicken Bake *recipe (page 148), I'm not a big fan of the "shake-in-a-bag" meat-coating mixtures. However, I like the idea. I simply think homemade tastes better. This version is simple and terrific on chops!*

161

GOLDEN FISH STICKS

Hey, you! Drop that box with the fisherman on it! You'll want to try these first. I guarantee that you'll never go back to frozen again.

2 ½ cups cornflakes
2 teaspoons lemon zest
½ teaspoon dried parsley flakes
2 tablespoons butter, melted
1 teaspoon lemon juice
Salt and freshly ground black pepper
1 pound fish fillets, cut crosswise into
    1-inch strips

Position rack to top third of oven and preheat to 500 degrees.
Grind cornflakes in a blender or food processor until it forms coarse crumbs.
Transfer to a mixing bowl; blend in lemon zest and parsley.
In a separate bowl, blend butter and lemon juice.
Season fish with salt and pepper.
Brush fish with lemon butter; dip into cornflake mixture.
Arrange fish on a baking sheet and bake approximately 10 minutes or until cooked through.

More Bearable Meal Suggestions
Serve with your favorite tarter sauce, fries, and *Cheesy Cheddar Biscuits* (page 89).

FLAMIN' JIFFY STEAK

This has romantic dinner for two written all over it. Inspired by the classic Steak au Poivre recipe, your date or significant other will be impressed by your mastery of fire!

2 beef cube steaks
1 teaspoon freshly ground black pepper
2 tablespoons butter
Salt
2 tablespoons brandy

Sprinkle cube steaks on both sides with pepper, pressing in firmly with hands.
In a medium skillet over medium-high heat, brown steaks in butter for 1 minute on each side or until desired doneness.
Sprinkle with salt.
Add brandy to skillet; carefully flame.
Remove steaks to a warm serving platter.
Pour pan drippings over steaks.

Use Your Tool
Although this is not necessarily regarded as a "dangerous" cooking technique, use caution when lighting brandy.

More Bearable Meal Suggestions
Serve with au gratin potatoes and a green salad followed by a delicious slice of *Kentucky Pie* (page 213) for dessert.

163

COCONUT CHICKEN

Can't escape to a tropical island anytime soon? Me neither. That's why I try to escape at home as often as possible.

1 ½ pounds boneless skinless chicken breasts, cut into thirds
2 teaspoons each salt and sugar
2 ½ cups flaked coconut
1 cup cornstarch
¼ cup all-purpose flour
2 eggs, beaten
¼ cup water
Vegetable oil for deep frying

Sprinkle chicken pieces with salt and sugar; place in a sealed container and refrigerate for at least 4 hours.
Remove from container; rinse lightly and pat dry.
Using deep fryer or deep skillet, heat oil.
In a mixing bowl, combine coconut, cornstarch, and flour; mix well.
In a separate bowl, blend eggs and water.
Place chicken into egg wash; then coat thoroughly in coconut mixture.
Carefully place chicken into hot oil and cook until golden brown.
Remove and dry on paper towels.

Use Your Tool
Do not crowd chicken pieces in oil while frying and watch the oil so it doesn't burn. Remember to always use caution when cooking with hot oil.

More Bearable Meal Suggestions
In keeping with an "island" theme, sauté some pineapple pieces in a little butter until thoroughly heated and serve on the side along with your favorite dipping sauces and hot cooked rice.

TEXAS BARBEQUE SHRIMP KABOBS

I love barbequed seafood, especially shrimp. This is one of my favorite ways to eat it (preferably with a few Texan Bears at the table). Yee-ha!

2 pounds jumbo shrimp, pealed, deveined
16 slices bacon, cut in half
1 large yellow or white onion,
    cut into 1-inch squares
1 each red and green bell pepper,
    cut into 1-inch squares
24 cherry tomatoes
24 fresh medium-sized mushrooms
1 cup barbeque sauce
8 (16-inch) metal skewers

Preheat oven to 400 degrees.
Wrap each shrimp with a half slice of bacon and skewer;
follow with alternating vegetables.
Place each loaded skewer on a baking sheet; bake
for 6-8 minutes.
Remove from oven and baste with sauce; return to
oven for 6-8 minutes.

More Bearable Meal Suggestions

Baste with *Your Own Barbeque Sauce* (page 93). Serve
over rice or with a side of *Potato Salad from Hell* (page
99).

165

BBQ HASH-BURGERS

Roast beef hash is a tasty alternative to ground beef that few people remember to try. If you haven't, this is a simple and tangy mouth-watering introduction.

1 (15-ounce) can roast beef hash
½ cup red onion, finely chopped
½ cup green bell pepper, finely chopped
⅓ cup barbeque sauce
2 teaspoons cider vinegar
1 tablespoon Worcestershire sauce
Salt and freshly ground black pepper to taste

In a large skillet over medium-high heat, brown hash, onion, and green bell pepper.
Add barbeque sauce, vinegar, and Worcestershire sauce; heat until warmed thoroughly; stirring occasionally.
Season with salt and pepper.
Spoon mixture onto hamburger buns.
If desired, top with slices of Swiss cheese.

More Bearable Meal Suggestions
Naturally, serve with French fries and preferably with beer or *Sweet Iced Tea* (page 71).

2 pounds chuck roast, cubed
3 tablespoons vegetable oil
1 small yellow or white onion, chopped
1 cup fresh mushrooms, sliced
1 (10-ounce) can tomato soup
¼ cup beef broth
1 bay leaf
¼ teaspoon freshly ground black pepper
1 teaspoon paprika
¼ pint sour cream

In a large skillet over medium-high heat, brown beef on all sides; place meat in slow cooker.
In the same skillet over medium heat, sauté onion and mushrooms for only 2-3 minutes; add to slow cooker with remaining ingredients—except sour cream.
Blend ingredients in cooker; cook at a LOW setting for 8-9 hours.
Serve mixture over noodles or rice and top with a spoonful of sour cream.

BEEF GOULASH

The name has never sounded to me as appetizing as its taste. Frankly, I don't care after one delicious bite. This is another great reason to use your slow cooker.

167

A TRIBUTE
Meat Broth

Special attention and recognition need to be given to meat broth, whether derived from hoof, beak, or snout.

Broth has been used in cooking for centuries as a tried-and-true base for soups, sauces, and gravies. Today, more residential and professional kitchens are discovering and exploring its amazing versatility.

I was lucky. My family used broth in the kitchen consistently. I even recall my parents adding it to our dog's dry food when she grew tired of it. Of course, we had to add broth to all of her meals after that!

Add flavor to rice, potatoes, or pasta by boiling them in broth instead of water; baste with broth to retain moisture when roasting meats; and revive leftover meats, casseroles, and side dishes by adding broth.

I recommend keeping a can of broth in your kitchen pantry at all times.

SALMON CRACKER CAKES

It just makes sense that a Bear cookbook would have salmon in it. Wouldn't you agree?

1 pound skinless salmon fillet,
    cut into ½-inch pieces
1 cup crushed Ritz crackers
    (or any butter-flavored cracker)
    (saltine crackers optional)
¼ cup red onion, finely chopped
¼ cup heavy cream
1 tablespoon dried dill
1 tablespoon yellow mustard
1 egg
½ teaspoon each minced garlic,
    Tabasco sauce, salt
½ cup all-purpose flour
2 tablespoons vegetable or olive oil

Preheat oven to 350 degrees.
In a large bowl, combine all ingredients except flour and oil; blend well.
Gently form mixture (easiest with clean hands) into 4 patties, measuring a half cup per patty.
Spread flour onto plate or clean surface; lightly coat both sides of each patty with flour.
Heat oil in a large skillet over medium heat.
Add patties and cook for 10 minutes or until golden brown, turning once.
Transfer patties to baking sheet and bake for 10 minutes.

More Bearable Meal Suggestions
Serve on buns with lettuce and tomato, or serve with *Hush Puppies* (page 80) and *Tangy Coleslaw* (page 96).

170

POT ROAST, MY SWEET

My affinity for pot roast goes back to childhood. It's one of my favorite comfort foods. I love to experiment with the classic recipe. This stovetop variation is so damn good that it hurts!

¼ cup all-purpose flour
1 teaspoon salt
¼ teaspoon freshly ground black pepper
4 pounds chuck roast
2 tablespoons vegetable oil
2 cups water
½ teaspoon each dried parsley flakes,
     oregano, celery seed
½ cup frozen concentrated orange juice
4 sweet potatoes, peeled and halved
4 yellow or white onions, sliced

In a small bowl, blend 1 cup water, parsley, oregano, and celery seed; set aside.
In a shallow dish, combine flour, salt, and pepper.
Place roast in dish and coat all sides with flour mixture.
In a stockpot over medium heat, brown roast on all sides in oil.
Pour herb and water mixture over roast; reduce heat to simmer (low).
Cover and cook for 2 hours.
In a small bowl, combine remaining cup of water and frozen concentrated orange juice; pour juice mixture over roast.
Add potatoes and onions.
Cover and continue simmering on low heat for 1 hour or until roast and vegetables are tender.
Transfer roast and vegetables to serving platter.

PORK CHOPS IN ORANGE-TOMATO SAUCE

Dress up boring pork chops with this tasty recipe! It's perfect for entertaining.

8 center-cut pork chops
1 cup all-purpose flour
½ cup seasoned bread crumbs
2 teaspoons salt
1 tablespoon paprika
Dash of freshly ground black pepper
Pinch of thyme
3-4 tablespoons vegetable oil

Sauce

2 cups tomato sauce
1 cup orange juice
2 tablespoons each honey, vegetable oil
1 tablespoon brown sugar

Preheat oven to 350 degrees.
In a shallow dish, combine flour, bread crumbs, and seasoning.
Place chops in dish and coat all sides with flour mixture.
In a skillet over medium heat, quickly brown each chop on both sides in oil; transfer each to a baking sheet.
In a saucepan over low heat, combine sauce ingredients and simmer for 3-4 minutes.
Pour 1-1 ½ cups of sauce over chops and bake for approximately 75 minutes.
Pour remaining sauce over rice when serving chops.

SOUTHWEST GRILLED CHICKEN

Keep that fire extinguisher handy because this chicken marinade is H-O-T, hot!

1 medium jalepeño pepper, diced
1 (4-ounce) jar pimientos with liquid
½ cup lime juice from concentrate
½ cup peanut oil
¼ cup pepper sauce
1 tablespoon cayenne pepper
½ tablespoon salt
3 large cloves garlic, minced
4 (6-ounce) boneless skinless chicken breasts

Combine all ingredients except garlic and chicken into a blender; process until smooth.
Stir in minced garlic.
Add chicken to a resealable plastic bag; pour marinade over chicken.
Remove majority of air in bag and seal; refrigerate 4 hours or overnight.
Remove chicken from bag and discard marinade.
Grill chicken over medium-hot fire for 10-15 minutes or until meat is thoroughly cooked.
Garnish with sour cream and serve with fresh salsa. Try *Love My Salsa* (page 18).

More Bearable Meal Suggestions
Serve with lightly grilled flatbread and your favorite beer and *New Classic Sangria* (page 19).

173

BOARD-WALK DOGS

Referring to this as a regular "chili dog" is a complete underestimation. I consider this a "super industrial strength" dog— the only kind that can satisfy a Bear's appetite.

1 ½ pounds lean ground beef
1 medium red onion, finely chopped
½ medium green bell pepper, chopped
1 (15-ounce) can tomato sauce
⅓ cup ketchup
1 teaspoon each chili powder, salt
½ teaspoon each garlic powder,
      freshly ground black pepper
8 each favorite brand of hot dogs and buns

Recommended Toppings

1 large white onion, chopped
1 large tomato, chopped
1-2 cups extra-sharp cheddar cheese, shredded

In a large skillet over medium-high heat, brown ground beef, red onion, and bell pepper until meat is cooked thoroughly.
Drain; return beef mixture to skillet.
Add tomato sauce, ketchup, chili powder, salt, garlic powder, and black pepper to skillet; simmer for 10 minutes.
Meanwhile, cook hot dogs according to package directions.
Place dogs on buns; spoon sauce over dogs and add desired amounts of toppings.

More Bearable Meal Suggestions

What else? BEER.

1 cup dry red or white wine
1 small yellow or white onion, grated
2 tablespoons soy sauce
1 teaspoon dried parsley flakes
2 cloves garlic, minced
1 large yellow or white onion, sliced
Brisket

Preheat oven to 325 degrees.
Combine wine, soy sauce, parsley, garlic, and grated onion in a deep roasting pan or large Dutch oven.
Add brisket; turn to coat both sides.
Sprinkle top and sides of brisket with sliced onions.
Cover top of pan with aluminum foil; place lid on top.
Cook for approximately 3 hours or until tender.
Place brisket on cutting board and let cool for at least 15 minutes.
Reserve juice and onions from pan.
Thinly slice brisket across the grain and place on serving platter.
Pour reserved juice and onions over brisket before serving.

More Bearable Meal Suggestions

Serve with *Yam-pple Casserole* (page 73) and *Sage Dressing* (page 84).

EASY NAPA BRISKET

Cooking brisket is very Zen-like for me—the ritual of the preparation, the patience needed while baking. For anyone who has never tried it, I assure you it's worth the wait. One bite will confirm that.

175

The Bird, the whole Bird, and nothing but the Bird

For those of you who have always been afraid of attempting the time-honored tradition of cooking the holiday turkey, fear no more. It is much simpler than you may suspect thanks to our friends at the U.S. Department of Agriculture (USDA).

According to the USDA, today's standard turkeys are younger, tenderer, and cook more quickly. The USDA recommends a new set of cooking and thawing times instead of those found in older cookbooks or reference guides. Cook at 325 degrees and always use a meat thermometer. Whe completely cooked, the temperature in the innermost part of the thigh should read 180 degrees, while the internal temperature of the breast should read 170 degrees. For more information, go online to www.fsis.usda.gov or call (800) 535-4555.

Roasting Times

Raw Weight	Unstuffed	Stuffed
8 to 12 pounds	2 ³/₄ to 3 hours	3 to 3 ¹/₂ hours
12 to 14 pounds	3 to 3 ³/₄ hours	3 ¹/₂ to 4 hours
14 to 18 pounds	3 ³/₄ to 4 ¹/₄ hours	4 to 4 ¹/₄ hours
18 to 20 pounds	4 ¹/₄ to 4 ¹/₂ hours	4 ¹/₄ to 4 ³/₄ hours
20 to 24 pounds	4 ¹/₂ to 5 hours	4 ³/₄ to 5 ¹/₄ hours

Thawing Times

No matter the thawing method, always place turkey on a tray or pan to contain any juices. If you choose the cold-water method, make certain to change the water every 30 minutes. Never thaw the turkey on a kitchen counter at room temperature, according to the USDA.

Thawing Time in Refrigerator
(24 Hours per 4 pounds)

8 to 12 pounds	1 to 2 days
12 to 16 pounds	2 to 3 days
16 to 20 pounds	3 to 4 days
20 to 24 pounds	4 to 5 days

Thawing Time in Cold Water
(30 minutes per pound)

8 to 12 pounds	4 to 6 hours
12 to 16 pounds	6 to 8 hours
16 to 20 pounds	8 to 10 hours
20 to 24 pounds	10 to 12 hours

SIMPLE ASIAN STEAK

4 (5-ounce) sirloin steaks
4 teaspoons soybean oil
6 large mushrooms, thickly sliced
½ cup green onion, chopped
Salt and freshly ground black pepper to taste

Place steaks on a broiler pan lightly coated with
nonstick cooking oil spray.
Broil steaks briefly on each side maintaining a rare
doneness in center.
Remove meat from broiler and place on a cutting
board; cube meat.
Heat oil in a large skillet over medium heat.
Add meat and mushrooms; sauté until desired
doneness.
Add green onions during last minute of cooking.
Salt and pepper to taste.

Use Your Tool
If you must cook steaks one at a time, remember to
distribute mushrooms and green onions evenly.

More Bearable Meal Suggestions
This recipe allows you tremendous freedom when
serving—so have fun! Serve with mashed potatoes and
gravy, or pour over rice and your favorite Chinese take-
out noodles. Or try dipping them in sauces as a snack.

You heard me. I said "simple"! For some cooks, this may be too simple. However, I included it for two reasons: I love the taste, and I wanted this to remind readers that steak isn't just for grilling in the backyard. You can (and should) enjoy this anytime without thinking it's too much fuss to make.

177

SWEET & SOUR PORK SPARERIBS

See if you can make these better than the take-out place. I bet you can.

¼ cup peanut oil
3 pounds spareribs, cut into 1-inch pieces
¼ cup green bell pepper, finely chopped
¼ cup red onion, finely chopped
1 teaspoon minced garlic
¾ cup pineapple juice
¾ cup water
¾ cup cider vinegar
3 tablespoons ketchup
½ teaspoon Worcestershire sauce
1 tablespoon soy sauce
½ cup brown sugar
2 tablespoons cornstarch

Heat oil in a large skillet until hot.
Add spareribs and brown for several minutes.
Remove from skillet; set aside.
Reserve 2 tablespoons of oil in skillet; add green pepper, onion, and garlic and brown for several minutes. Add ½ cup of pineapple juice, water, vinegar, ketchup, Worcestershire sauce, soy sauce, and brown sugar; stir well.
In a small bowl, blend remaining ¼ cup of pineapple juice and cornstarch into a smooth paste.
Add this to skillet and stir well.
Bring skillet to a boil; add spareribs and reduce heat to simmer.
Cook uncovered, stirring often, for approximately 1 hour or until meat is tender.

More Bearable Meal Suggestions
Serve alone as a snack or with *Quick Mushroom Risotto* (page 98).

178

2 pounds bratwurst links
1 quart dark beer
1 cup water (optional)

In a 2-3 quart stockpot, add brats to beer and bring to a boil.
Reduce heat, cover and simmer for 15 minutes.
Remove brats and grill (or brown in a large skillet) for an additional 15 minutes.
Serve on bratwurst-style buns with a spicy mustard.

Use Your Tool

Use any kind of beer for boiling; however, the darker beers are recommended for a more distinct flavor.

ULTIMATE BEAR MEAT

Why the name? When you combine beer and red meat together, it should be obvious. This is quintessential Bear food.

BOURBON BEEF TENDERLOIN

This preferred cut of meat deserves a remarkable marinade. It's a marriage made in grilling heaven.

1 cup bourbon
2 cups water
1 cup brown sugar
2/3 cup soy sauce
1 small bunch fresh cilantro, chopped
1/2 cup lemon juice
1 tablespoon Worcestershire sauce
1 teaspoon dried parsley flakes
3 sprigs fresh thyme, chopped
1 (4-5 pound) beef tenderloin, well trimmed

In a mixing bowl, combine bourbon, brown sugar, soy sauce, cilantro, lemon juice, Worcestershire sauce, parsley, and thyme.
Add tenderloin to shallow dish; pour marinade over meat and cover.
Refrigerate for 10-12 hours, turning meat several times.
Prepare grill.
Drain marinade into saucepan; heat to a boil.
Reduce heat and cook for 3-4 minutes.
Place meat on oiled grill.
Cook with lid closed, turning often and basting with cooked marinade for 45 minutes or until a meat thermometer inserted in the thickest part registers 135 degrees (medium rare).
Remove from heat and let stand 10 minutes before slicing.

More Bearable Meal Suggestions
Serve with your favorite kind of potatoes. I prefer fluffy, buttery mashed potatoes.

½ cup taco sauce
¼ cup Dijon mustard
2 tablespoons lime juice
½ teaspoon dried parsley flakes
6 fresh boneless skinless chicken breast halves
2 tablespoons butter
6 tablespoons plain yogurt
6 slices of lime

In a large bowl, combine taco sauce, mustard, lime juice, and parsley.
Add chicken and marinate at room temperature for at least 30 minutes.
Remove chicken and reserve marinade.
In a large skillet over medium heat, melt butter; add chicken and cook for 3-5 minutes on each side until lightly browned.
Add marinade and cook approximately 5-7 minutes or until fork tender, turning chicken once or twice.
Remove chicken and place into warm oven.
Turn heat to higher setting and boil marinade for an additional 1-2 minutes.
When serving individual portions, pour marinade over chicken and top with a tablespoon of yogurt and a slice of lime.

Use Your Tool

For food safety, it is important to boil the marinade before serving. If you prefer to marinate before you grill or broil the chicken, always discard unused marinade.

More Bearable Meal Suggestions

Serve with Spanish or brown rice, *Love My Salsa* (page 18) and *New Classic Sangria* (page 19) or beer.

CHICKEN BREASTS WITH YOGURT AND LIME

Here's another great way to jazz up those chicken breasts you like to eat during the week because they're so convenient to have around.

181

LIVER AND ONIONS DIJON

As my family members would attest, I was a peculiar child. In fact, I held a deep dark secret from my childhood friends and my siblings—I always loved when my mother served liver and onions. This is a variation of my mother's preparation as a tribute to the secret she and I shared.

½ medium yellow onion, thinly sliced
½ cup green onion, chopped
2 tablespoons butter
¼ cup all-purpose flour
¼ teaspoon each: salt and freshly ground
      black pepper
1 pound calf's liver, cut ¾-inch thick (4 slices)
1 tablespoon water
2 teaspoons Dijon mustard
2 teaspoons Worcestershire sauce
1 teaspoon lemon juice

In a large skillet over medium heat, melt 1 tablespoon of butter; add yellow onion and cook until tender and lightly golden brown.
Add green onion and cook 1–2 minutes only.
Remove onions from skillet; set aside.
In a shallow bowl, combine flour, salt, and pepper.
Coat each liver slice thoroughly in flour mixture.
In the same skillet, melt remaining butter over medium heat; cook each liver slice for 3 minutes on each side.
Transfer liver to warm oven or platter.
Return onion mixture to skillet; add water, mustard, Worcestershire, and lemon juice; cook and stir until thoroughly heated.
Spoon onion mixture over liver and serve.

More Bearable Meal Suggestions
Serve with *Triple Cheese Mac Attack* (page 68) and beer.

LEMON DILL FISH STEAKS

Once again, cooking with mayonnaise rules supreme! This time it's a wonderful sauce for fish.

³/₄ cup mayonnaise
2 teaspoons dried dill weed, crumbled
1 tablespoon lemon juice
1 tomato, seeded and diced
Dash of freshly ground black pepper
4 (6-ounces each) salmon or halibut steaks,
     cut ³/₄-inch thick

In a mixing bowl, combine mayonnaise, dill, and lemon juice.
Transfer ½ cup of mixture into another small bowl; set aside.
Add tomato and pepper into first mixture and blend well.
Brush fish steaks with the ½ cup mayonnaise mixture
(without tomato).
Grill or broil steaks 6 inches from heat, turning once,
approximately 7-8 minutes or until fish flakes easily.
Serve with tomato-mayo mixture.

Use Your Tool
You may substitute 2 tablespoons of chopped fresh dill
for the dried variety.

More Bearable Meal Suggestions
Serve with *Southern-Baked Corn* (page 77) and *Hush Puppies* (page 80).

183

Section VI

WAY BEYOND THE HONEYPOT

PACKING
THE FUDGE

*There's nothing
more rewarding
than giving the
gift of sweets,
especially
something
delicious like
fudge. I love
whipping up a big
ol' batch of fudge,
wrapping the
pieces in waxed
paper and packing
them into gift
boxes. In fact, it's
much more fun
to invite friends
over (experience
not required) and
have a good old-
fashioned fudge
packing party.*

3 cups (18 ounces) chocolate chips,
        semisweet or milk chocolate
1 (14-ounce) can sweetened condensed milk
Dash of salt
1 ½ teaspoons vanilla extract
1 cup chopped walnuts

In a medium saucepan over low heat, melt chocolate
chips with sweetened condensed milk and salt.
Remove from heat; stir in vanilla and nuts.
Spread mixture into a greased, foil-lined 8 or 9-inch
square pan.
Refrigerate for at least 2 hours or until firm.
Turn fudge onto cutting board; peel foil away and cut
into squares.
Store covered in refrigerator.

Use Your Tool
For thinner fudge pieces, line a baking sheet with waxed
paper and spread fudge evenly on top before chilling.

More Bearable Meal Suggestions
Enough said.

1 cup (2 sticks) butter
½ cup sugar
½ cup brown sugar
1 egg
2 tablespoons orange juice
1 tablespoon orange zest
2 ¾ cups all-purpose flour
1 teaspoon baking soda
½ cup walnuts, chopped

In a large bowl, beat butter and sugar (preferably with an electric mixer) until light, approximately 3 minutes. Beat in brown sugar (medium speed); add egg, orange juice, and zest.
Stir in flour and baking soda; followed by nuts.
Shape mixture into two 9 x 3-inch bars; wrap in plastic. Refrigerate for 8 hours.
Heat oven to 350 degrees; cut bars into thin slices.
Place slices on lightly greased baking sheets; bake for approximately 10 minutes or until edges begin to brown.
Cool on wire racks.

Use Your Tool
You can attempt the difficult task of making this dough by hand; however, using an electric mixer is easier and offers the best result.

ORANGE NUT COOKIES

It's hard to find a store-bought cookie this unique. The trick to its one-of-a-kind flavor is in the dough's refrigeration time. These are well worth the wait.

187

BALI HAI
PIE

*I swear that I
would be willing
to be stranded on
a deserted island
for the rest of my
life if all I had to
eat was this.*

¹/₃ cup butter
1 ½ cups sugar
3 eggs
2 tablespoons all-purpose flour
2 tablespoons cornmeal
1 cup crushed pineapple, drained
1 cup sweetened coconut, shredded
½ cup pecans, chopped
1 teaspoon vanilla extract
1 unbaked 9-inch pie shell

Preheat oven to 350 degrees.
In a large bowl (preferably using an electric mixer),
beat butter, sugar, eggs, flour, and cornmeal until well
blended.
Add pineapple, coconut, nuts, and vanilla; mix well.
Pour mixture into pie shell.
Bake approximately 45 minutes or until wooden
toothpick inserted into center comes out almost clean.

188

1 stick (1/2 cup) butter, softened
1 cup sugar
3 eggs, beaten
1 cup all-purpose flour
1/2 teaspoon baking powder
1/4 teaspoon salt
1/2 teaspoon nutmeg
1/4 cup milk
2 tablespoons molasses
1/4 teaspoon baking soda
1 pound yellow seedless raisins
2 cups walnuts, chopped
1/4 cup whiskey

Preheat oven to 300 degrees.
In a large mixing bowl, cream together the butter and sugar; add beaten egg.
In a separate bowl, blend flour, baking powder, salt, and nutmeg; add to butter mixture; blend.
In a third bowl, blend baking soda into the molasses; add to the butter/flour mixture; blend.
Add raisins, nuts, and whiskey and mix well.
Pour into a greased and floured loaf pan and bake for 2 hours.

DAMN GOOD WHISKEY CAKE

This was inspired by a dearly departed friend who once made a deal with himself when he gave up drinking. Whenever we sat together at the pub, he would order a shot of milk, shoot it back and exclaim, "Hoo-wee! That's damn good whiskey!" He would have loved this cake.

EASY CROCK-POT BREAD PUDDING

*Why so easy?
Other than the
prep time for some
of the ingredients,
the crock does
all of the work.
This is especially
convenient when
you need your
oven and stovetop
free for preparing
other meal items.*

8 slices raisin bread, diced
2 green apples, peeled, cored, and thinly sliced
3 eggs
1 cup sugar
1 teaspoon cinnamon
1/2 teaspoon nutmeg
2 cups half-and-half
1/4 cup bourbon or brandy
1 cup pecans, chopped
1/4 cup butter, melted

Place bread in a greased 3-quart or larger Crock-Pot;
blend in apples.
In a separate bowl, beat eggs with sugar, cinnamon, and
nutmeg.
Add half-and-half and bourbon or brandy; mix well.
Sprinkle bread with pecans.
Pour egg mixture over bread mixture; drizzle butter on
top.
Cover and cook at LOW setting for 3-4 hours or until
apples are tender and custard is set.
Turn off heat and let stand for 15 minutes.

More Bearable Meal Suggestions
This is a great warm and comforting wintertime dessert,
and is best served with vanilla ice cream.

190

1 package white cake mix
1/2 cup pecans, chopped
3/4 cup butter, melted
1 (21-ounce) can apple pie filling

Preheat oven to 350 degrees.
In a mixing bowl, blend melted butter into cake mix;
add nuts and mix well until crumbled (and a bit sticky)
consistency.
Spread pie filling into a 9 x 13-inch baking dish.
Sprinkle cake mixture evenly over filling. Bake for 18-
20 minutes or until filling begins to bubble.

More Bearable Meal Suggestions

Best served warm over vanilla ice cream.

SIMPLE
APPLE
CRISP

*It's a great "last-
minute" dessert.
In fact, I believe
its convenience
ranks it as one of
the best examples
of comfort dessert
food. Find comfort
in this anytime!*

NUTTER SQUARES

*WARNING: It's so
easy that you'll be
enjoying this all
of the time. You'll
begin to wonder
if it's a dessert
or a snack.*

1 cup light corn syrup
1 cup sugar
1 (12-ounce) jar crunchy-style peanut butter
6 cups crisp rice cereal
½ cup pecans, chopped

In a 2-quart microwave-safe mixing bowl, combine
syrup and sugar; heat in a microwave oven on HIGH
setting for 2 minutes or until sugar is dissolved.
Stir in peanut butter until well blended.
Add cereal and nuts; mix well.
Pour into a greased 9 x 13-inch dish or pan.
When cooled cut into squares.

Use Your Tool
Depending upon their size when cut, this recipe may
yield up to two dozen.

1 cup sugar
2 tablespoons all-purpose flour
1 teaspoon baking powder
3 Red Delicious apples, unpeeled, cored,
      and chopped
1 cup sliced almonds
1 egg

Preheat oven to 350 degrees.
In a mixing bowl, blend sugar, flour, and baking powder.
Add apples, nuts, and egg; blend well.
Pour into an 8-inch square baking dish.
Bake for 50 minutes.

EASY APPLE NUT PUDDING

Sometimes necessity really is the mother of invention. Case in point: In an excited yet foolish moment on the phone with a friend, I offered to host a very last-minute dinner party thinking that I had plenty of food in my kitchen. I didn't, and I had no time to shop. Miraculously, I fed my friends and found enough basic kitchen ingredients on hand to whip up a simple dessert. Lucky for me it was a hit.

193

ULTIMATE SNICKER COOKIES

*I would like
to take this
opportunity to
thank all of
the candy bar
manufacturers
for a lifetime of
snacking pleasure;
and an additional
"thank you"
for inventing
their bite-sized
versions. To show
my gratitude,
I offer this…*

½ cup sugar
½ cup brown sugar, firmly packed
½ cup butter, softened
½ cup creamy-style peanut butter
1 teaspoon vanilla extract
1 egg
1 ½ cups all-purpose flour
½ teaspoon each baking powder, baking soda
¼ teaspoon salt
10 Snickers "Fun Size" (bite-sized) candy bars

Preheat oven to 375 degrees.
In a large bowl, combine sugar, brown sugar, butter,
peanut butter, vanilla, and egg; beat well.
Add flour, baking powder, baking soda, and salt to
sugar mixture; mix well.
Form approximately ⅓ cup of dough smoothly and
completely around each candy bar.
Place each 4 inches apart on ungreased baking sheets.
Bake for 15 minutes or until golden brown.
Let cool for 10 minutes; transfer to wire racks to cool
completely.

1 (18-ounce) box chocolate cake mix
1 cup mayonnaise
1 cup water
3 eggs

Preheat oven to 350 degrees.
Grease and lightly flour two 9-inch round cakes pans;
set aside.
In a large bowl, using electric mixer, blend all
ingredients on a low speed; then beat at a medium
speed for 2 minutes.
Pour into prepared cake pans.
Bake for 30 minutes or until centers of cakes spring
back to the touch.
Cool in pans for 10 minutes on wire racks.
Remove from pans and cool completely.
Serve with the following choices: sprinkled with
confectioners' sugar, topped with frosting, or my
favorite, smothered in chocolate syrup.

Use Your Tool
WARNING: Mixing ingredients by hand will not offer
best results—and it's exhausting!

CLASSIC NO-FUSS CHOCOLATE CAKE

*Many readers may
already be familiar
with this recipe—
hence the term
"classic." However,
I felt compelled to
include it because
it's so damn easy
to make and has a
wonderful flavor…
Life should always
be filled with
chocolate cake.*

A TRIBUTE
Choco-Logic

My parents knew from the time I was very young that I was a handful. Once, I explained to my mother in a very rational tone that I decided to forego the peanut butter and jelly sandwich she had made me for lunch with a "nutritious" (and very large) chocolate bar. She offered the sandwich with a stern reprimand and a deal: If I could prove that the chocolate was healthier than the sandwich, I would have permission to eat it.

And so my stubborn quest began.

I returned to my parents at dinner with my research and began with some sort of boastful preamble like, "The cocoa bean has been cultivated and prepared through the centuries for a variety of consumptions—some medicinal, some recreational…" My pathetic yet polished oration dragged on as they reluctantly listened through every course and patiently sat through my "closing arguments," some of which follows:

Chocolate is derived from coca beans, and beans are considered vegetables. Sugar is derived from either sugar cane or sugar beets—both are in the "vegetable category." Hence, ladies and gentlemen, chocolate is a vegetable.

To make matters worse, I went one step further to proclaim that the milk chocolate variety contained dairy, which qualified candy bars technically as a "health food." I also concluded that chocolate-covered raisins, strawberries, cherries, and orange slices were all still considered "fruit", so I should be allowed to eat as much as I wanted.

Well, I never got the candy bar. However, I did find a chocolate Kiss on my pillow that night.

PEANUT BUTTER BANANA CRUNCH

I've always had a thing for peanut butter and banana sandwiches, so I thought I'd take it a step further...

4 cups (6 medium-sized) bananas
1 tablespoon lemon juice
½ teaspoon ground cinnamon
½ cup all-purpose flour
½ cup packed brown sugar
3 tablespoons butter
⅓ cup chunky-style peanut butter

Preheat oven to 375 degrees.
In an 8 x 8-inch baking pan or dish, combine peeled bananas, lemon juice, and cinnamon; stir until bananas are well-coated.
In a mixing bowl, combine flour, brown sugar, butter, and peanut butter until mixture becomes crumbly. Sprinkle the peanut butter mixture evenly over the bananas.
Bake for 25 minutes.
Serve warm; top with whipped cream if desired, or serve with vanilla ice cream.

Use Your Tool
A 9-inch pie pan or dish may be substituted.

1 9-inch ready-made piecrust
1 cup sugar
1 ½ teaspoons ground cinnamon
½ teaspoon ground cloves
½ teaspoon allspice
½ teaspoon ground nutmeg
½ teaspoon ground ginger
½ teaspoon salt
2 eggs
1 (12-ounce) can evaporated milk
1 ½ cups canned pumpkin
    (NOT "pumpkin pie mix")

Preheat oven to 425 degrees.
Place piecrust carefully into a 9-inch pie pan.
In a mixing bowl, preferably using an electric or hand mixer, blend all remaining ingredients until smooth.
Pour into pie shell; bake for 15 minutes.
Reduce heat to 350 degrees and continue baking for 35-40 minutes.

GRANDMA'S PUMPKIN PIE

Simple. Wholesome. And it's one of my favorite pies ever! You just can't go wrong with this one.

ANYTIME
TIRAMISU

*For the record,
I'll admit that I
can be a tiramisu
snob when
dining at Italian
restaurants. It is
only because I've
enjoyed authentic
Italian versions
and believe me,
once you've tasted
the "real thing,"
there is no settling
for mediocre.
However, when I
love a particular
recipe and want it
anytime, I like a
version that's fast
and delicious.*

1 (8-ounce) package cream cheese
½ cup confectioners' sugar
2 tablespoons dark rum
2 cups of whipped topping
1 package ladyfinger biscuits
½ cup dark roast coffee, finely ground
5 tablespoons cocoa powder

In a large mixing bowl, cream together cream cheese
and sugar.
Add rum and blend well.
Fold in the whipped topping; set aside.
In the bottom of a 9 x 13-inch dish, arrange half of the
package of ladyfingers in a single layer.
Carefully sprinkle coffee over each ladyfinger biscuit.
Spread half of cream cheese mixture evenly over top.
Add additional layer of ladyfingers; top this layer with
remaining cream cheese mixture.
Top with an even coat of cocoa powder using a sifter or
spoon.
Refrigerate for 15-20 minutes before serving.

2 sticks (1 cup) butter
4 squares unsweetened chocolate
2 cups sugar
4 eggs
1 cup all-purpose flour
2 teaspoons vanilla extract

Preheat oven to 350 degrees.
In a large saucepan over medium-low heat, melt
· butter and chocolate squares; stirring constantly until
completely melted.
Remove from heat and add remaining ingredients.
Stir until well blended; pour into an ungreased 9 x 13-
inch baking dish.
Bake for 20 minutes.

Use Your Tool
If you prefer your brownies "less gooey," bake for an
additional 5 minutes.

INTENSE BROWNIES

*Turn off your
computer and the
TV, unplug the
phone, turn down
the lights, and
sit back because
these bad boys
are intense from
the very first bite.
Relax and enjoy!*

BUTTER-
RUM
SUNDAE

If you ever get tired of pouring the same old squeeze-bottle syrups over your ice cream, try this instead. It is especially impressive for summertime entertaining.

¼ cup butter
1 package creamy white frosting mix
        (for 2-layer cake)
2 tablespoons light corn syrup
⅓ cup evaporated milk
¼ cup each chopped pecans, walnuts
¼ cup rum
Vanilla ice cream

In a saucepan over medium-high heat, melt butter, stirring until browned; remove from heat.
Add half of the frosting mix and corn syrup; stir to blend.
Add remaining frosting mix; stir until well-blended.
Gradually stir in evaporated milk.
Return to medium heat; cook and stir until heated through.
Remove from heat; stir in nuts and rum.
Serve over ice cream.

1 package lemon cake mix (with pudding)
2 cups whipped topping, thawed
1 egg
confectioners' sugar

Preheat oven to 350 degrees.
In a large mixing bowl, combine cake mix, whipped topping, and egg.
Using an electric mixer on medium speed, beat ingredients until blended.
The batter's texture should be very thick and sticky.
Drop batter by teaspoonfuls into sugar and roll lightly to coat.
Place each piece of coated batter several inches apart on an ungreased baking sheet.
Bake for 10-12 minutes or until lightly brown.

LEMON SNOW-FLAKES

This is a terrific holiday gift recipe when you don't have the time to race around shopping for multiple gifts. One batch makes approximately 5 to 6 dozen. Your friends and family will love receiving these!

CASHEW
PUDDING

*I must confess.
I'm addicted to
cashews. I picked
up the habit from
my father—the
king of cashew
connoisseurs. This
one's for you, dad!*

½ cup raw cashews
2 cups whole milk
½ cup honey
2 tablespoons butter
Dash of salt
4 teaspoons cornstarch
½ cup sweetened coconut, shredded

Combine all ingredients (except coconut) in a blender
until smooth and well-blended.
Pour mixture into saucepan and cook on very low heat;
stirring constantly until a thickened pudding texture
forms.
Remove from heat; pour pudding evenly into dessert
bowls or cups.
Refrigerate for at least 1 hour or until pudding hardens.
Before serving, sprinkle tops of servings with coconut.

Use Your Tool
If burning the pudding is a concern, use a double boiler
instead of a saucepan as an extra precaution.

1 box yellow cake mix
1 (3-ounce) package instant lemon pudding
4 eggs
³/₄ cup vegetable oil
1 ¼ cups 7-UP soft drink

Topping
2 eggs, beaten
1 tablespoon all-purpose flour
1 stick (½ cup) butter
½ cup sugar
1 cup canned crushed pineapple, drained
3 tablespoons lemon juice

Preheat oven to 350 degrees.
In a mixing bowl, beat together cake mix, pudding, eggs, and oil; add 7-UP.
Pour mixture into a greased 9 x 13-inch baking pan.
Bake for 40 minutes or until lightly browned or a wooden toothpick inserted into center comes out clean.
Remove from oven and let cool for 15 minutes on rack.

For topping
In a nonaluminum saucepan over medium heat, combine all topping ingredients (except lemon juice); cook uncovered for 10 minutes, stirring often.
Remove from heat and stir in lemon juice.
Spread topping over cake and serve.

CLASSIC 7-UP CAKE

My love for food began very early in life. It was apparent when my first-grade teacher, Mrs. Edmonds, would bring this cake to class on special occasions, and I was the only student to ask for the recipe.

MANDARIN ORANGE PIE

This is an incredibly easy way to use canned fruit when you need an impressive dessert in a hurry.

1 (8-ounce) package cream cheese, softened
¼ cup orange marmalade
2 cups frozen whipped topping, thawed
1 cup pecans, finely chopped
2 (15-ounce) cans mandarin orange slices,
    well-drained
1 (9-inch) ready-to-use graham cracker
    piecrust

In a mixing bowl, blend cream cheese and marmalade;
stir in whipped topping.
Place oranges in a single layer on bottom of pie crust;
sprinkle evenly with nuts.
Spoon cheese mixture over oranges.
Top cheese mixture with a layer of remaining oranges.
Cover and refrigerate for at least 2 hours or overnight.

Use Your Tool
This dessert can be made more decorative if you arrange
orange pieces in a circular pattern.

2/3 cup sugar
5 fluid ounces (1/4 pint) water
1/2 cup cocoa powder

In a saucepan over high heat, combine sugar and water;
bring to a boil, stirring until sugar is dissolved.
Remove from heat and whisk in cocoa powder.
Bring mixture just to a boil; remove from heat again
and let stand and thicken as it cools.
Pour into a sealed container and refrigerate unused
portion.

Use Your Tool
For extra fun and flavor, try adding a dash of cinnamon
or a shot of Grand Marnier when whisking the cocoa
powder.

More Bearable Meal Suggestions
Let's face it. The serving possibilities are endless. Don't
just pour this over ice cream. Set your imagination free!

BEST HOMEMADE CHOCOLATE SAUCE

*I believe that life
gives you few
guarantees—
death, taxes,
and the fact
that everything
tastes better with
chocolate sauce
poured over it.
That, for sure, is
a guarantee.*

A Little Lowdown on the Sweet Stuff

Has your brown sugar hardened?

Place a slice of bread and the brown sugar in a sealed bag and the sugar should soften in a few hours.

Sticky business

Before measuring syrup or honey, oil the cup with cooking spray. Afterward, rinse with hot water.

Meringue migraine

The trick to slicing a meringue pie cleanly is to keep a tall container of hot water nearby. Dip your cutting knife into the water between every cut to keep the meringue from sticking.

It's just too wet!

To avoid making a pie too juicy, sprinkle 2 tablespoons of flour and 1 tablespoon of sugar onto the bottom of the pie crust before adding the pie filling. This helps with absorption and flavor.

Or, not wet enough...

Want more juice from your lemons or limes? Heat them in the microwave for 30 seconds after first squeezing, then squeeze again!

ROCKY ROAD SQUARES

1 (21-ounce) package fudge brownie mix
Vegetable oil (per package directions)
Egg(s) (per package directions)
½ cup evaporated milk
1 cup semisweet chocolate chips
2 cups miniature marshmallows
1 cup pecans, chopped

Preheat oven according to brownie mix instructions.
Grease a 9 x 13-inch baking dish; set aside.
Prepare brownie mix per package directions using
vegetable oil, egg, and substituting evaporated milk for
water.
Spread mixture into dish and bake per package
directions—DO NOT OVERBAKE.
Remove from oven; sprinkle top with chocolate chips.
Let stand 5 minutes; spread melted chips evenly over
brownies.
Sprinkle top evenly with marshmallows and nuts.
Bake an additional 3-5 minutes or just until
marshmallows begin to melt.
Cool for 30 minutes before slicing and serving.

*Need an
impressive yet easy
dessert to bring
to a party? This
is a great way to
"doctor up" a basic
box of brownie
mix and make
it appear that
it's made from
scratch. I've tried
this many times—
it works.*

209

NOODLE PUDDING

Hard-core recipe collectors will appreciate this: I've had this recipe for so many years that I can't remember who gave it to me. I wish I did so I could thank them again. I love it and make it regularly because it's so delicious and comforting.

1 (16-ounce) container sour cream
1 (8-ounce) container cottage cheese
6 eggs
1/2 cup brown sugar
1/2 teaspoon salt
4 ounces egg noodles
1/2 cup raisins
Ground cinnamon to taste
1/4 cup (1/2 stick) butter

Preheat oven to 400 degrees.
In a large saucepan, boil noodles according to package directions.
Meanwhile, in a mixing bowl using an electric mixer on MEDIUM-LOW speed, beat sour cream and cottage cheese adding one egg at a time.
Reduce mixer speed to LOW and add sugar and salt.
Drain noodles completely; stir in noodles and raisins to cheese mixture until well-blended.
Pour mixture into a greased 9 x 13-inch baking dish.
Sprinkle top with cinnamon and dot with butter.
Bake for 30 minutes; then reduce heat to 325 degrees and bake an additional 30 minutes.

1 ¼ cups all-purpose flour
½ teaspoon baking soda
½ teaspoon salt
1 stick (½ cup) butter softened
        (room temperature)
¼ cup brown sugar
¼ cup sugar
½ teaspoon vanilla extract
1 egg
1 ½ cups chocolate chips
1 cup pecans or walnuts, chopped

Preheat oven to 325 degrees.
In a bowl, combine flour, baking soda, and salt; set aside.
In a mixing bowl using an electric mixer on a MEDIUM
speed, beat butter and both sugars until creamy.
Reduce mixer to LOW speed and beat in flour mixture.
Stir in chocolate chips and nuts.
Drop heaping teaspoons of batter onto ungreased
baking sheets approximately 2 inches apart.
Bake for 12-15 minutes or until edges become golden brown.
Remove from oven and cool completely on wire racks.

ALL-AMERICAN CHOCOLATE CHIP COOKIES

Just in case you did not already have a recipe for one of the most popular cookies ever invented— here it is. I have many, but this is one of my favorites. Enjoy!

211

CARROT CAKE

*Some "store-
bought" cakes
are perfectly
acceptable for
many occasions—
or at least that's
what you want
to believe when
you're in a hurry.
This, on the other
hand, is one of the
best homemade
cakes I've ever
made. It will
always remind
you that there is
a difference with
homemade.*

2 cups packed light brown sugar
½ cup each vegetable oil, buttermilk
¼ cup honey
3 eggs
2 cups all-purpose flour
2 teaspoons ground cinnamon
1 teaspoon baking soda
½ teaspoon each salt, nutmeg
2 cups grated carrots
½ cup each raisins, chopped pecans

Frosting
1 (8-ounce) package cream cheese
1 stick (½ cup) butter, softened
2 cups confectioners' sugar, sifted
2 teaspoons vanilla extract

Preheat oven to 350 degrees.
In a mixing bowl using an electric mixer on MEDIUM
speed, cream together brown sugar, oil, buttermilk,
honey, and eggs.
In a separate bowl, sift together flour, cinnamon,
baking soda, salt, and nutmeg.
Stir into egg mixture.
Stir in carrots, raisins, and nuts until well-combined; set
aside.
Grease and flour a 9 x 13-inch baking pan.
Pour batter evenly into pan; smooth with spatula.
Bake 30-35 minutes or until toothpick inserted into
center of cake comes out clean.
Cool on wire rack.
Meanwhile, cream together cream cheese and butter in
a mixing bowl.
Add sugar and vanilla; beat until light and fluffy.
Frost cake.

1 9-inch ready-made piecrust
2 eggs
1 stick (½ cup) butter, melted
⅓ cup all-purpose flour
1 cup dark chocolate chips
1 cup pecans, chopped
1 cup sugar
1 tablespoon bourbon
⅛ teaspoon salt

Preheat oven to 325 degrees.
Form piecrust into 9-inch pie pan; set aside.
Add eggs to a blender or food processor and blend until frothy.
Add remaining ingredients and mix just until chocolate chips become coarsely chopped.
Pour mixture into pie shell and bake for 50-60 minutes or until center rises and crust becomes golden brown.

Use Your Tool
Take a shot of bourbon for yourself once you pop that pie into the oven as an extra "at-a-boy" for a job well done!

KENTUCKY PIE

You must know by now that I love to cook with liquor. Well, here's another way to incorporate bourbon into your diet. Is that so wrong?

213

RUM RAISIN DESSERT

*If you thought
ricotta cheese
was only good
for lasagna,
think again.*

3 tablespoons raisins
¼ cup light rum
1 pound fresh ricotta cheese
¼ cup confectioners' sugar
Grated peel of 1 lemon
⅓ cup walnuts, chopped

In a small shallow dish or bowl, soak raisins in rum for
1 hour.
In a mixing bowl, beat ricotta with a fork until softened.
Stir in remaining ingredients including raisins and rum.
Spoon mixture into dessert glasses and refrigerate for at
least 1 hour.

Use Your Tool
Slice an extra lemon to use as a garnish.

1 ¾ cups all-purpose flour
1 teaspoon baking powder
½ teaspoon salt
1 ½ cups packed light brown sugar
1 stick (½ cup) butter, melted
2 eggs
2 teaspoons vanilla extract
1 (12-ounce) package dark chocolate chips
1 cup walnuts, coarsely chopped

Preheat oven to 325 degrees.
In a medium-sized bowl, sift together flour, baking powder, and salt; set aside.
In a large bowl, combine sugar and butter; stir well.
Add eggs and vanilla; stir until smooth.
Add flour mixture; stir until just blended.
Stir in chocolate chips.
Spoon batter into a greased 8-inch baking pan.
Sprinkle top evenly with nuts.
Bake for 40 minutes or until toothpick inserted into center comes out slightly moist.
Cool thoroughly on wire rack.
Cut into squares.

EASY CHOCOLATE CHIP BLONDIES

Once had a love, and he was a Bear. Soon found out, he lacked body hair… so, I drown my sorrows in a giant pan of blondies and a gallon of milk. Then, I brushed off the crumbs and decided to keep dating despite the obstacles.

APPLE CHEDDAR COOKIES

*Although I love
the classic variety
of cookie, I must
admit, I enjoy the
more inventive
ones—like this!
It makes you
think "outside the
cookie box."*

1 stick (½ cup) butter, softened
½ cup sugar
1 egg
1 teaspoon vanilla extract
1 ½ cups unbleached flour
½ teaspoon baking soda
½ teaspoon ground cinnamon
½ teaspoon salt
6 ounces, extra-sharp cheddar cheese, shredded
1 ½ cups Red Delicious apples, cored,
      peeled, and chopped
¼ cup chopped pecans

Preheat oven to 375 degrees.
In a mixing bowl preferably using an electric mixer,
cream together butter and sugar until light and fluffy.
Stir in egg and vanilla.
Add dry ingredients; blend well.
Stir in cheese, apples, and nuts.
Drop rounded teaspoonfuls of batter onto ungreased
baking sheet 3 inches apart.
Bake for 15 minutes.

216

1 (14-ounce can) sweetened condensed milk
²/₃ cup chocolate syrup
2 cups whipping cream
½ cup milk chocolate chips

In a large bowl, blend sweetened condensed milk and chocolate syrup.
In a mixing bowl preferably using an electric or hand mixer, whip cream until stiff.
Fold whipped cream into chocolate mixture.
Lightly stir in chocolate chips.
Line a 9 x 5-inch loaf pan with aluminum foil.
Pour mixture into pan and cover with foil.
Freeze for at least 6 hours or until firm.

MILK CHOCOLATE ICE CREAM

Homemade ice cream is easier than you think. This requires no expensive gourmet ice cream machines, no rock salt, and no hand cranks...try it!

217

GLAZED
APPLE
CAKE

*There is nothing
quite like the fresh
taste of apples
in a cake. I warn
you—this one is
very hard to share.*

1 ³/₄ cups sugar
1 cup vegetable oil
3 eggs
2 cups all-purpose flour
1 teaspoon salt
1 teaspoon baking soda
1 teaspoon ground cinnamon
¹/₂ teaspoon nutmeg
1 cup pecans, finely chopped
3 cups Granny Smith apples, cored,
        peeled, and thinly sliced

Glaze
2 tablespoons butter, melted
1 teaspoon vanilla extract
1 ¹/₂ cups confectioners' sugar
hot water

Preheat oven to 350 degrees.
In a mixing bowl, beat sugar, oil, and eggs together.
In a separate bowl, sift together flour, salt, baking soda,
cinnamon, and nutmeg.
Stir flour mixture into sugar mixture; blend well.
Fold in nuts and apples.
Pour batter into a greased 9 x 13-inch baking pan.
Bake for 50-60 minutes.
Meanwhile, to make the glaze, blend melted butter,
vanilla, and confectioners' sugar.
Stir in hot water carefully—just enough to allow glaze
to become a sauce or thin syrup consistency.
Remove cake from oven and drizzle glaze over cake.
Glaze will harden once it cools.

6 cups frosted cornflakes cereal
(crushed into 2 ¼ cups)
1 stick (½ cup) butter, melted
2 (8-ounce each) packages cream cheese,
softened
1 (7-ounce) jar marshmallow crème or "fluff"
1 cup walnuts, chopped
⅔ cups sliced fruit, fresh or canned
(apricots, bananas, Mandarin oranges,
peaches, berries, kiwi, pineapple)

SUMMER FRUIT PIZZA

Don't panic! The title may imply a "light and fluffy" dessert, but it's not. This is definitely made for Bear appetites.

Preheat oven to 325 degrees.
In a large bowl, blend crushed cereal flakes and butter.
Spread mixture onto a 12-inch pizza pan.
Press mixture firmly with back of spoon.
Bake for 7-8 minutes or until a light golden brown.
Remove from oven and cool completely.
In a separate bowl, combine cream cheese and marshmallow crème or "fluff."
Spread evenly over crust.
Sprinkle nuts evenly over cream cheese mixture.
Arrange fruit slices on top; refrigerate until thoroughly chilled.
Cut into 12 slices.

Use Your Tool
Top each slice with a spoonful of whipped topping for added effect.

BUTTER-SCOTCH NUT BARS

*There is something
wonderfully
comforting
about the flavor
of butterscotch.
It's like a big
Bear hug. I've
been a fan since
my first piece
of butterscotch
candy. This dessert
always reminds me
of my childhood
and of home.*

3/4 cup butter
1 ½ cups packed brown sugar
1 tablespoon vanilla extract
2 eggs
2 cups all-purpose flour
1 ½ teaspoons baking powder
1 cup walnuts, chopped
1 cup butterscotch-flavored chips

Preheat oven to 350 degrees.
In a medium-sized saucepan over low heat, melt butter
and brown sugar, stirring occasionally.
Remove from heat when butter has melted completely.
In a large mixing bowl using an electric mixer at
MEDIUM speed, beat together sugar mixture, vanilla,
and eggs.
Add flour and baking powder; beat until well-blended.
Stir in ½ cup each of nuts and butterscotch chips.
Pour mixture into a greased 9 x 13-inch baking pan.
Sprinkle top with remaining nuts and chips. Bake for
25-30 minutes or until top is golden brown and center
is set.
Cool completely, approximately 1 hour.

ALMOND-OATMEAL COOKIES

½ cup sugar
½ cup packed brown sugar
½ cup butter, softened
½ teaspoon vanilla extract
1 egg
1 ½ cup quick-style oats
½ cup all-purpose flour
½ cup ground toasted almonds
½ teaspoon baking soda
¼ teaspoon baking powder
⅛ teaspoon each: salt, cinnamon
½ cup sliced almonds

Preheat oven 375 degrees.
In a large mixing bowl, preferably using an electric or hand mixer, combine both sugars, butter, vanilla, and egg.
Stir in oats, flour, ground almonds, baking soda, baking powder, salt, and cinnamon; combine well.
Stir in sliced almonds.
Drop rounded teaspoons of dough 2 inches apart on ungreased baking sheets.
Bake approximately 10 minutes or until golden brown.
Cool for at least 5 minutes.
Remove from baking sheets and serve.
Makes approximately 3 dozen.

*"There once was a Bear from Nantucket... ."
No, seriously, there was. His name was Coby and this was his favorite kind of cookie. I wonder if he still lives there... .
So, where was I?
Oh, yes, "with a cookie so large, he could dunk it."*

221

FROSTED COLA CAKE

This is for all of the cola junkies of the world. When you just can't drink enough, start cooking with it!

1 stick (½ cup) butter
1 cup dark cola
1 square dark baking chocolate
¾ cup miniature marshmallows
½ cup each shortening, vegetable oil
2 teaspoons vanilla extract
2 cups sugar
2 eggs
¾ cup buttermilk
2 ⅓ cups all-purpose flour
¾ cup cocoa powder
1 teaspoon baking soda
1 ½ teaspoons baking powder
½ teaspoon salt

Frosting

1 stick (½ cup) butter, softened
½ cup cocoa powder
1 teaspoon vanilla extract
¼ cup each dark cola, chocolate syrup
3 cups confectioners' sugar

Preheat oven to 350 degrees.
Bring butter, cola, and chocolate square to a boil in a saucepan.
Remove from heat; blend in marshmallows until completely melted.
Set aside and cool to room temperature.
In a large mixing bowl, combine shortening, oil, vanilla, and sugar; add eggs and buttermilk and beat until smooth.
In a separate bowl, blend flour, cocoa, baking soda, baking powder, and salt.
Blend half of flour mixture into shortening mixture; add cola mixture and blend well.

222

Add remaining flour mixture and beat several minutes until smooth.
Pour thick batter into a greased and floured 9 x 13-inch baking pan.
Bake for 35-40 minutes.

To Prepare Frosting

Cream butter, cocoa, vanilla, cola, and chocolate syrup in a mixing bowl.
Mix in confectioners' sugar one cup at a time; beat until smooth.
If frosting seems too dry, add more cola and beat until smooth.
Frost cake when warm.

GERMAN CHOCOLATE CHEWS

Although these are chocolate-free, I like the name because its flavor reminds me of German chocolate cake frosting— my favorite!

³/₄ cup (1 ½ sticks) butter
3 tablespoons sugar
1 cup all-purpose flour
2 ¼ cups packed dark brown sugar
3 eggs; separated
½ cup each chopped pecans, chopped walnuts
³/₄ cup sweetened coconut flakes

Preheat oven to 350 degrees.
In a mixing bowl preferably using an electric mixer, beat butter and sugar at a MEDIUM speed; stir in flour.
Pat the thick dough into a greased 9 x 13-inch baking pan.
Bake 13-15 minutes or until edges are a light brown. Remove from oven; set aside.
In the same mixing bowl, beat brown sugar and egg yolks at LOW speed.
Add nuts and coconut; mix and set aside.
In a separate bowl, beat egg whites several minutes until frothy (not stiff).
Gently fold egg whites into coconut mixture; spread over baked flour mixture.
Bake for 30 minutes.
Cool approximately 1 hour on wire rack before cutting into squares.

Use Your Tool
Here's another recipe that can be accomplished without an electric mixer, but it's not recommended—especially when beating the egg whites.

224

1 (15-ounce) can pears, chopped
        (juice reserved)
1 (6-ounce) package lime-flavored gelatin
1 (8-ounce) package cream cheese, softened
1 pint whipping cream

Drain pears into a measuring bowl; add enough water to juice to make 2 cups.
Transfer juice mixture to a saucepan and heat to a boil.
Remove from heat and pour into blender.
Add gelatin and cream cheese; blend thoroughly.
Chop pears and place into a 9 x 13-inch dish.
Pour gelatin mixture over pears; refrigerate for 1 hour or until mixture has slightly thickened.
In a separate bowl, whip cream until thick; fold into pear mixture.
Pour mixture into mold and refrigerate for at least 4 hours or overnight.

LIME GELATIN MOLD

Believe it or not, after publishing Bear Cookin', *Bears frequently asked me why I didn't include a gelatin mold. Many of them remember growing up with these delicious desserts. So, here is one of my favorites.*

BEE-ELZA-BUTTER COOKIES

*I warn you.
You'll get a little
naughty with
these sinful treats.
They are pure
E-VIL. (The devil
made me do it.)*

1 package devil's food cake mix
2 cups peanut butter chips
1 (16-ounce) container sour cream

Preheat oven to 350 degrees.
In a mixing bowl, combine all ingredients; blend well.
Drop rounded teaspoonfuls of batter at least 3 inches
apart on greased baking sheets.
Bake for 9–11 minutes.
Cool on wire racks.

Use Your Tool
An electric mixer is preferred but not necessary.

1 ¼ cups butter, softened
2 cups sugar
6 eggs
½ cup creamy peanut butter
2 cups all-purpose flour
¼ cup peanuts, chopped

Preheat oven to 350 degrees.
In a mixing bowl, preferably using an electric mixer, cream butter and sugar; beat until fluffy.
Add eggs one at a time while continually beating.
Mix in peanut butter followed by flour.
Pour mixture into two greased 9 x 5-inch loaf pans.
Bake for 55-60 minutes or until toothpick inserted into center comes out clean.
After 20 minutes of baking, sprinkle nuts evenly over top of each loaf and continue baking.

PEANUT BUTTER POUND CAKES

I refer to my friend, Jerome, as "Peanut Butter Bear" because of our ongoing rivalry over who has the most and best peanut butter recipes. This is the only recipe that we agree is our favorite. It has now become a traditional holiday ritual to give one loaf to each other and keep the second loaf all to ourselves. Thanks, P.B.!

CHAMELEON CAKE

*Is it a cobbler,
a pie or a cake?
One bite of this
exquisite dessert
and you won't
care what it
looks like. You'll
just know that
it's amazingly
delicious.*

6 large green apples, peeled, cored,
    and thinly sliced
2 cups plus 2 tablespoons sugar
2 teaspoons ground cinnamon
2 tablespoons plus 3 cups all-purpose flour
2 tablespoons lemon juice
2 ½ teaspoons baking powder
½ teaspoon salt
2 large eggs
1 stick (½ cup) butter, melted
½ cup water
1 teaspoon vanilla extract

Preheat oven to 350 degrees.
In a large mixing bowl, add apples, 1 cup of sugar,
cinnamon, 2 tablespoons flour, and lemon juice; mix
well and set aside.
In a second bowl, add 1 cup sugar, baking powder, salt,
and 3 cups flour; mix well.
In a third bowl, whisk together eggs, butter, water, and
vanilla.
Add egg mixture to dry ingredients; blend to create
thick dough.
Spread 1 ½ cups of dough over bottom of a greased
9 x 13-inch baking pan.
Pour apple mixture over dough.
Drop tablespoons of remaining dough evenly over
apples.
Sprinkle with remaining tablespoons of sugar.
Cover pan with aluminum foil and bake for 30 minutes.
Remove foil and bake an additional 45-50 minutes or
until golden brown.
Let pan cool on wire rack for at least 30 minutes.
Serve warm with or without ice cream.

228

²/₃ cup sugar
½ cup packed brown sugar
½ cup butter, softened
2 tablespoons whole milk
2 eggs
¾ cup all-purpose flour
½ teaspoon baking powder
¼ teaspoon salt
¼ cup creamy peanut butter
⅓ cup cocoa powder
½ cup each chocolate chips,
        peanut butter chips

Preheat oven to 350 degrees.
In a mixing bowl, combine sugars, butter, milk, and
eggs; mix well.
Add flour, baking powder, and salt; blend well.
Divide batter evenly into two separate bowls.
In one bowl of batter, blend in peanut butter and
peanut butter chips.
In the other bowl, blend in cocoa powder and chocolate
chips.
Spoon each batter, checkerboard-style, into a greased
9 x 9-inch baking pan.
Using a knife, gently swirl batters into each other.
Bake for 30-35 minutes or until toothpick inserted into
center comes out clean.
Cool for 1 hour before cutting into squares.

PEANUT BUTTER CUP BROWNIES

One of the most classic and most beloved dessert flavor combinations, chocolate and peanut butter, come together in a perfect brownie. It's a beautiful thing.

229

PINEAPPLE SURPRISE

What's the surprise? You'll be surprised at how easy it is to make. And your friends will be even more surprised that you made it.

2 large (15-20 ounce) cans crushed pineapple, drained
$^2/_3$ cup pineapple juice (reserved from canned pineapple)
3 teaspoons cornstarch
4 eggs, beaten
2 cups sugar
2 tablespoons butter
1 teaspoon cinnamon

Preheat oven to 350 degrees.
In a small bowl, blend pineapple juice and cornstarch into a loose paste.
Add pineapple, eggs, and sugar; mix well.
Pour mixture into a greased 9 x 13-inch baking dish.
Dot top with butter and sprinkle with cinnamon.
Bake for 1 hour.
Serve warm.

Recipe Terminology

al denté: Comes from Italian meaning "to the teeth." Describes foods, especially pasta, cooked soft enough to eat but not over overcooked.

appetizer: A small portion of food served before a meal.

arrowroot: Tasteless, starchy powder used as a thickening agent when combined with water.

au gratin: A casserole topped with cheese, butter, and bread crumbs and browned.

baste: To moisten foods during cooking with pan drippings or a flavorful sauce.

batter: Uncooked mixture usually made of flour, egg, and liquid and used as a base for most baked goods. Also can be used to dip foods before frying.

beat: To make mixture smooth by briskly whipping or stirring with a spoon, hand or electric mixer.

blanche: Process in which food is briefly immersed into boiling water then immediately soaked in cold water.

blend: To thoroughly mix two or more ingredients until smooth and uniform.

braise: Cooking technique usually applied on the stovetop using little fat to brown, then covering and cooking until done. Best suited for tougher meats.

broil: To cook food directly under or over heat source.

brown: To cook food quickly at a moderately high heat to brown the surface.

casserole: Name of a baking dish; used to describe recipes that include a combination of foods such as meats, vegetables, and grains that are bound by a thick sauce.

cayenne: A small, hot, chile pepper.

chop: To cut into small pieces.

chunk: To cut into bite-sized pieces usually 1-inch or larger.

231

clarified butter: Melted, unsalted butter skimmed of milk solids.

compote: Combination of cooked and fresh fruits served hot or cold.

cream: To beat with a spoon or electric mixer until smooth, light, and fluffy.

cube: To cut into cubes usually ½-inch to 1-inch in size.

cut in: To mix shortening with dry ingredients using a pastry blender or knives.

dash: Less than 1/8 teaspoon.

deep-fry: To cook in hot fat deep enough to completely cover food.

dice: To cut into cubes approximately ⅛ to ¾-inch size.

dollop: Can refer to a spoonful of soft food or a dash of a liquid.

dot: To scatter bits of an ingredient (usually butter) evenly over the surface of another food.

dough: A mixture of flour and liquid that is too thick to pour but thick enough to work with hands.

dredge: To coat food with a dry mixture—usually flour or bread crumbs.

drippings: Juices or liquefied fats left in pan after cooking meats or other food.

drizzle: To pour a thin mixture like melted butter or icing over food in a fine stream.

egg wash: Egg whites or yolks usually mixed with small amount or water then brushed onto baked goods to give color and sheen.

extract: Concentrated flavors from various foods, usually derived from evaporation or distillation; may be in solid or liquid form.

flake: To use a fork or other utensil to lightly break off pieces of food.

flank steak: A long, fibrous cut of beef from an animal's lower hindquarters and usually tenderized, marinated, and grilled or broiled.

flour: (verb) To lightly sprinkle or coat with flour.

fold: To gently turn (usually with a spatula) a lighter mixture into a heavier mixture several times until combined.

fry: To cook food in fat over moderate to high heat.

garnish: To decorate food or the dish on which food is served.

glaze: A thin, glossy coating applied to the surface of food.

grate: To cut food into small shreds usually with a food grater.

grease: (verb) To spread fat on a pan or cooking utensil to prevent food from sticking to surface.

grill: To cook on a rack directly over hot coals or other heat source.

grind: To reduce food into small particles using a variety of devices such as a food processor, blender, peppermill, mortar and pestle, or meat grinder.

half-and-half: A mixture of half cream and half milk.

hash: A dish of chopped potatoes, meat, and vegetables; often made by using leftovers.

hollandaise sauce: An emulsified sauce made from egg yolks and butter, usually flavored with lemon juice or vinegar.

hors d'oeuvre: Small, bite-sized foods occasionally served before a meal.

hushpuppy: A small, fried cornmeal dumpling. Legend has it that the name came from Civil War-era cooks who would feed their pet dogs with scraps of fried batter to quiet them.

icing: Also known as frosting; sweet mixture used to top or coat baked goods.

Italian seasoning: A blend of dried herbs, usually includes oregano, basil, thyme, rosemary, marjoram, red pepper, and sage.

jalapeño: A small, hot chile pepper, usually green.

jam: A thick, cooked mixture of fruit and sugar. Sometimes pectin is added.

jelly: A clear, cooked mixture of fruit juice and sugar. Sometimes pectin is added.

julienne: To cut food lengthwise into thin, matchstick strips approximately $1/8$-inch thick.

knead: To work dough with the heel of the hand using a pressing, folding motion.

kosher: Food prepared according to Jewish dietary laws.

lard: Rendered and clarified pork fat.

liqueur: A sweet alcoholic drink frequently used in baked desserts or dessert sauces and occasionally served with or after meals.

macerate: To soak fruit or other food in liquid in order to soften or flavor it. Brandy is commonly used.

mandoline: A compact, hand-operated slicing and cutting machine.

marinade: A seasoned liquid used to flavor and/or tenderize meat or other foods.

marinate: To soak food in a seasoned liquid to tenderize and flavor.

marmalade: A citrus fruit mixture, similar to jam, that includes the fruit peel.

marzipan: A sweet paste made from ground almonds, sugar, and egg whites.

mash: To crush food until smooth and evenly textured.

melt: To heat food until it becomes liquid.

meringue: A mixture of beaten egg whites and sugar.

mince: To finely chop food into small pieces, usually 1/8-inch or smaller.

mix: To blend ingredients.

mold: To form a food into a shape and chill or cook until firm enough for food to hold shape.

mull: To flavor a beverage by heating it with spices or other flavorings.

partially set: To chill gelatin until the consistency of egg whites.

phyllo: A Greek pastry mainly used in Greek dishes made of thin layers of dough; sometimes spelled "filo."

pinch: A small amount of a dry ingredient—approximately 1/16 of a teaspoon—usually enough to hold between the tips of the thumb and forefinger.

poach: To cook food in liquid at or just below the boiling point.

preheat: To allow an oven or pan to reach a specific temperature before adding food to be cooked.

reduce: To boil a liquid until a portion of it has evaporated.

render: To extract the fat from meat by cooking over low heat, then remove meat particles.

roast: To cook food in an uncovered pan in an oven using no added liquid.

sauté: To cook food quickly in a pan on the top of stove until browned.

scallion: Also known as a "green onion."

sear: To brown meat quickly over high heat using a broiler or stovetop skillet.

shallot: An onionlike flavored bulb related to the onion and garlic.

shred: To cut food into narrow strips; usually finer than julienne or grated methods.

sift: To pass dry ingredients through a mesh sifter; used to incorporate air and help make ingredients lighter.

simmer: To cook liquid at approximately 185 degrees, or just below a boil when tiny bubbles just begin to break the surface.

skim: To remove a substance from the surface of a liquid.

sliver: To cut food into thin strips or pieces.

soft peaks: To beat egg whites or whipping cream until peaks form when beaters are lifted, but tips curl over.

springform pan: A round cake pan commonly used to make cheesecakes and usually deeper than a standard cake pan. It has a clamp on its side that releases the sides from the bottom, leaving the cake intact.

steep: To soak in order to extract flavor or soften.

stiff peaks: To beat egg whites until peaks stand up straight when beaters are lifted, but are still moist and glossy.

strain: To pour liquid through a strainer or colander to remove solid particles.

whip: To beat ingredients usually by using a whisk or an electric mixer in order to incorporate air into a mixture and change its texture.

wok: A pan with a round bottom used to stir-fry foods.

Worcestershire sauce: This thin, dark condiment, originally bottled in Worcester, England, is used to season meat, sauces, gravies, and other various dishes.

yam: Sometimes thought of as and used like a sweet potato. However, it is from a different plant species and usually grown and eaten in South and Central America, parts of Asia, and Africa. True yams can be found in Latin American markets.

zest: The thin, brightly colored outer skin of a citrus fruit (not the white part).

Recipe Equivalencies

Dry Ingredients

½ tablespoon	=	1 ½ teaspoons
1 tablespoon	=	3 teaspoons
2 tablespoons	=	⅛ cup
4 tablespoons	=	¼ cup
5 tablespoons plus 1 teaspoon	=	⅓ cup
8 tablespoons	=	½ cup
12 tablespoons	=	¾ cup
16 tablespoons	=	1 cup
2 cups	=	1 pint
2 pints	=	1 quart
4 cups	=	1 quart
4 quarts	=	1 gallon

Liquids

1 tablespoon	=	½ ounce
2 tablespoons	=	1 ounce
4 tablespoons (or ¼ cup)	=	2 ounces
8 tablespoons (or ½ cup)	=	4 ounces
1 cup	=	8 ounces
2 cups	=	16 ounces or 1 pint
4 cups	=	32 ounces or 1 quart or 1 liter

Emergency Substitutions

Sometimes ingredient substitutions are necessary. Here are some common ones. Keep in mind that the recipe results may vary slightly.

Ingredient   Substitute

1 teaspoon baking powder............ 1/4 teaspoon baking soda
+ 1/2 teaspoon cream of tartar

1 cup honey.................................. 1 1/4 cup sugar + 1/4 cup liquid

1 cup oil...................................... 1/2 lb. butter

1 cup whole milk.......................... 1/2 cup evaporated milk plus 1/2 cup water

1 cup half-and-half...................... 1 cup evaporated milk,
or 7/8 cup milk + 3 tablespoons butter

1 cup buttermilk.......................... 1 tablespoon cider vinegar or lemon juice
+ enough milk to equal 1 cup and allow
to stand at least 5 minutes.

1 cup dairy sour cream................. 1 cup plain yogurt, or 1 cup evaporated milk
+ 1 tablespoon vinegar

1 tablespoon cornstarch............... 2 tablespoons flour,
or 1 1/3 tablespoons quick-cooking tapioca

1 tablespoon flour........................ 1/2 tablespoon cornstarch, or 2 egg yolks,
or 2 teaspoons quick-cooking tapioca.

1 tablespoon tapioca..................... 1 1/2 tablespoon all-purpose flour

1 oz. (1 square)
unsweetened chocolate................. 3 tablespoons cocoa + 1 tablespoon shortening

1 oz. (1 square)
semisweet chocolate.................... 1 square unsweetened + 1 tablespoon sugar

1 cup tomato sauce...................... 8-oz. can stewed tomatoes, blended,
or 3/4 cup tomato paste + 1/4 cup water.

1/2 lb. fresh mushrooms............... 4-oz. can mushrooms

Recipe Index

Subject Index